Gianfranco Marrone
Introduction to the Semiotics of the Text

Semiotics, Communication and Cognition

Edited by
Paul Cobley and Kalevi Kull

Volume 31

Gianfranco Marrone

Introduction to the Semiotics of the Text

—

ISBN 978-3-11-125942-0
e-ISBN (PDF) 978-3-11-068898-6
e-ISBN (EPUB) 978-3-11-068903-7
ISSN 1867-0873

Library of Congress Control Number: 2021942869

Bibliographic information published by the Deutsche Nationalbibliothek
The Deutsche Nationalbibliothek lists this publication in the Deutsche Nationalbibliografie;
detailed bibliographic data are available on the Internet at http://dnb.dnb.de.

Originally published in 2011 under the title
Introduzione alla semiotica del testo
Gius. Laterza & Figli
© Gianfranco Marrone

Translated by Alice Kilgariff

Copyright of the English edition:
© 2023 Walter de Gruyter GmbH, Berlin/Boston
This volume is text- and page-identical with the hardback published in 2022.
Cover design based on a design by Martin Zech, Bremen
Typesetting: Integra Software Services Pvt. Ltd.
Printing and binding: CPI books GmbH, Leck

www.degruyter.com

Foreword

This is the second of a number of books in the *Semiotics, Communication and Cognition* series which feature translations of important work in semiotics that are, as yet, little known in the world of Anglophone academia. Supported in a unique funding collaboration between de Gruyter and the International Association for Semiotic Studies (IASS), ostensibly these translations focus on work from the Greimassian tradition. Hugely influential, the works from this tradition have not been well served, in terms of numbers, by translations. This initiative, in addition to earlier translations in this series of books by Lotman, Sériot and Lagopoulos and Boklund-Lagopoulou, is designed to ameliorate that situation, to make available key works in English translation, to encourage cross-fertilization in semiotics and to promote further dialogue, learning and intellectual semiosis.

We are delighted to present Gianfranco Marrone's *Introduction to the Semiotics of Text*, to accompany José Enrique Finol's master work, *On the Corposphere*, published by de Gruyter Mouton earlier as part of this initiative. The question may arise, however, regarding why an 'introduction' should appear in a series such as *SCC*, dedicated to publishing books on high-level semiotic research. Moreover, the question might be asked why there is a need for a book on the idea of the 'text'. Not only is the idea of the 'text' almost naturalized in semiotics and beyond, but we also have in English the supreme tracking of the trajectory of that idea in Marrone's own summation, the magisterial volume entitled *The Invention of the Text* (Mimesis, 2014).

The answer to these questions is that the current book by Marrone is one of those rare phenomena in publishing: the 'introduction' which does not oversimplify but, at every instance, opens the door to new arguments, new theories and new perspectives on both the familiar and the unusual. From the very first sentence of the first chapter, where the reader finds themselves in the emotional homeliness of the Summer holidays, there is also an interruption in the manner of the best classical semiotics, defamiliarizing the entire situation. "Why not use an example that is more serious, institutional, socially recognised", asks Marrone in discussing the rationale for this opening, such as "a poem for instance, or a novel, a film, an advert, a legal text, a newspaper article, a comic, or even (as we will do later) a painting?" (p. 1). It is clear that this volume, in the manner of Barthes' rarely fulfilled 1970 call for 'semioclasm', sees no area beyond the reach of semiotics. Gianfranco Marrone is one of the most productive analysts, even among Italian semioticians, with books on the semiotics of the city, of food, of branding, of laziness, of newspapers, etc., and with only his aforementioned 2014 book on the text, alongside the volume co-edited with Dario Mangano, *Semiotics of Animals in Culture: Zoosemiotics 2.0* (Springer, 2018), as yet translated into English. The semioclasm in the current book is marked by a commitment to critique which constantly overturns not only the mainstream views of signification in society as evidenced in ideology and bland conformity but also in the limitations of unambitious forms of semiotics

Integral to this commitment is "a very general theoretical model known as the *generative path of meaning*" (p. 22). As Marrone puts it,

> According to this theoretical model, the meaning of any text is articulated through signification according to levels of pertinence placed at various depths, in order of complexity and tangibility. As such, the most profound levels are abstract and simple, whilst the superficial ones are instead more tangible and complex (p. 22).

The most profound level of signification is to be found in *narrative structures*, with their anthropomorphic and fundamental layers. However, there are also structures at the level of enunciation which are to be negotiated: *discursivity* and *textualization*. Upon these are built the "great basic semantic oppositions such as life/death, culture/nature, euphoria/dysphoria, and so on" (p. 25).

This complex model is intimately related to the much-vaunted work of Greimas and his associates which is regrettably so seldom available to Anglophone scholars. So, no doubt readers will wish to rush forward to Chapter 3 of this volume where they can find a coruscating discussion of Greimas' 'semiotic square'. However, we would discourage this. In order to follow the developing logic of Marrone's implementation of "a dialectic of the implicit and the explicit in generative terms" (p. 58) we would, instead, encourage readers to take this volume at a measured pace, paying due attention to the way in which the effortful sedimentation of layers of meaning might be uncovered bit by bit. The book certainly does not seek to close down analysis by presenting the Greimas model as an 'application' to be used in interrogating the 'semiosphere'; rather, it opens to a future to be apprehended by way of the generative path, Greimassian sociosemiotics and Lotmanian distinctions on the relation of texts to what is putatively 'non-text'.

Like all great introductions, this book facilitates further exploration by its readers, principally in the form of 'further reading' at the end of each chapter. Semiosis begets semiosis; but any reader of this volume will be aware that it is not unfocused and directionless. We are confident that readers will go away with a new appreciation and awareness of textuality and the role of the generative path in bringing about that appreciation and awareness.

As editors of this series, we would like to thank Gianfranco Marrone for his generous co-operation in the production of his own translation and expansion of his original volume, *Introduzione alla semiotica del testo*.

Paul Cobley and Kalevi Kull
June 2021

Contents

Foreword —— V

I	**Theoretical fundamentals —— 1**
I.1	Expression and content —— 1
I.2	Sender and receiver —— 2
I.3	Communication and signification —— 3
I.4	Inferences and culture —— 4
I.5	Difference and values —— 6
I.6	Relations and narrations —— 8
I.7	Form and substance —— 9
	Recommended bibliography —— 11
II	**Basic principles —— 13**
II.1	Historical emergences, plural filiations —— 13
II.2	How we recognise textuality —— 16
II.3	On the principle of pertinence —— 19
II.4	The generative stratagem —— 21
II.5	Text and culture —— 24
	Recommended bibliography —— 25
III	**The logics of narration —— 27**
III.1	Narration and narrativity —— 27
III.2	Elementary structures of signification —— 28
III.2.1	Statics: Fundamental relationships —— 29
III.2.2	Dynamics: Basic operations —— 32
III.2.3	Complex term and neutral term —— 34
III.2.4	The constitution of axiologies —— 35
III.3	Elements of narrative grammar —— 37
III.3.1	Subject and object —— 37
III.3.2	Subject of doing and subject of being —— 38
III.3.3	Objects and values —— 38
III.3.4	Actants and actors —— 39
III.4	Programmes, modalities, identities —— 40
III.4.1	Modalities and modelling —— 40
III.4.2	Basic programmes and instrumental programmes —— 42
III.4.3	The Addresser —— 43
III.5	Canonical narrative schema —— 45
III.5.1	Narration and reasoning —— 48

III.5.2	The narrative presupposition —— 48	
III.6	Polemics and strategies —— 49	
III.6.1	Conflicts and identity —— 50	
III.6.2	Alterity and simulacra —— 50	
III.6.3	Multiplication of subjectivities —— 51	
III.6.4	Tactics —— 52	
III.6.5	Meta-strategies and culture —— 52	
III.7	Logics of affect —— 53	
III.8	Canonical schema of passions —— 55	
III.9	Forms of life —— 58	
	Recommended bibliography —— 61	
IV	**Enunciation and discourse —— 62**	
IV.1	From communication to enunciation —— 62	
IV.2	Languages and subjectivity —— 64	
IV.2.1	Enunciation in philosophy —— 64	
IV.2.2	Enunciation in linguistics —— 65	
IV.2.3	Enunciation in semiotics —— 66	
IV.2.4	Enunciation and action —— 68	
IV.3	Efficiency and efficacy —— 70	
IV.4	Strategies of knowledge —— 71	
IV.5	Intertextuality, interdiscursivity, intermediality —— 75	
IV.6	Themes and figures —— 78	
	Recommended bibliography —— 82	
V	**Image, sensoriality, body —— 84**	
V.1	Forms and substances of expression —— 84	
V.2	Figurative and plastic —— 85	
V.2.1	Time in images —— 86	
V.2.2	Semi-symbolism —— 90	
V.3	Figurative reasoning —— 91	
V.3.1	Representation and perception —— 91	
V.3.2	Degrees of figurativity —— 93	
V.4	Plastic language —— 96	
V.4.1	Plastic categories —— 97	
V.4.2	Classique and baroque —— 98	
V.5	Sensory experience, corporeality, space —— 99	
V.6	Synaesthesia —— 102	

V.7	Space and subjectivity —— 107
V.8	The aesthetic grasp —— 111
V.9	A final glance —— 113
	Recommended bibliography —— 118

Appendix: A History of the notion of text —— 121

1	Text and society —— 121
2	Sociosemiotics without textuality? —— 124
3	From philology to linguistics —— 127
3.1	A brilliant utopia —— 127
3.2	A negative entity —— 128
4	Aesthetics and methodology —— 130
4.1	From Work to Text —— 130
4.2	Discourse and narrativity —— 132
5	Hermeneutics, deconstructionism, textualism —— 134
5.1	Contributions of hermeneutics to sociosemiotics —— 135
5.2	The outside-text —— 136
5.3	The hermeneutic arch —— 140
5.4	Interpretation and configuration —— 142
6	Semiotics of text —— 144
6.1	Interpretative cooperation —— 146
6.2	Cultures and genres —— 149
6.3	Generative path and projects of description —— 150
7	Basics of sociosemiotics —— 158
8	Ethnology and semiotics of culture —— 162
8.1	Structural anthropology —— 163
8.2	Interpretative anthropology —— 165
8.3	Cultural models —— 166
9	Farewell to representation —— 172
	Recommended bibliography —— 175

References —— 177

Index —— 195

I Theoretical fundamentals

I.1 Expression and content

It's the summer holidays. We arrive on an unfamiliar Mediterranean island in the morning. It is still too early for the hotel, so we rent a scooter and drive around to look for a beach – preferably one that is free and somewhat wild – where we can while away the hours with a refreshing swim and a little sunbathing. We travel the road that tracks the coast, but the high metal fences of a long series of villas block our view of the sea. Between bursts of bougainvillea and jasmine shrubs, signs appear every so often, advertising bars selling sandwiches and drinks, shops selling parasols, sun loungers and sunscreen, a news stand, a few B&Bs; and sharp turns leading to a little town that can be spotted atop a distant hill. There is no sign of a beach. Everything seems to look endlessly the same: the eyes get distracted, the other senses grow restless, any hope for the coveted swim is fast disappearing. Suddenly, the glare from an expanse of glittering metal announces the existence of a car park. Well, not a car park exactly but a series of cars that have been abandoned in an improvised manner at the side of the road, and thickening into a uniform metal mosaic next to an opening that seems to lead down towards the coast. A more careful look confirms what we had already suspected – that is the way to the sea: just a few minutes of trekking along that arduous path and there is the longed-for destination. The isolated beach is there with its crystalline water, golden sand and no telephone signal. We have found it, the pile of cars unexpectedly signalled it for us.

Here we have a banal event that functions as an ideal opening to a book on The Semiotics of the Text. An event much like any other that happens in everyday life. Of course, it could be said that this particular situation is somewhat unique as it is set in a time and a space that are not habitual. This is undoubtedly true. But what happens is absolutely ordinary: a series of signs that set in motion a mechanism of action and reaction that is at the same time obvious and complex, made up of tensions, perceptions, feelings, expectations, delusions, intuitions, considerations, interpretations and so on.

So why should we begin this book on the science of texts by recounting an episode that we can define as a fable? Why not use an example that is more serious, institutional, socially recognised – a poem for instance, or a novel, a film, an advert, a legal text, a newspaper article, a comic, or even (as we will do later) a painting?

There are many reasons. The first being that, in a situation of everyday life, we find ourselves faced with a clear sign, with all of the principal traits, conditions and consequences that generally tend to characterise signs. Thanks to a number of cars parked on the side of the road (or rather, thanks to the fact that we noticed them), we understood *that* was where the path leading to the beach must have been, and we found it. The vision of that number of cars creates the basis for what is called a *signifying expression*, the thing through which we were able to understand. The presence of the path leading down to the sea at that precise point of the road constitutes what we will call the *signified content*, or simply what we understood from the sight of all those parked cars. The union between the signifying expression (which is sensory, perceptual, empirical) and the signified content (that is instead intellectual, interpretative, cognitive) gives rise to what we call *sign*, which is that increase in knowledge in the island's geography that we had been lacking until that moment. Finally finding the path required to get to the sea created the *pragmatic effect* of the sign, which is in this case a positive effect, since the path leading to the beach was exactly what we were looking for.

I.2 Sender and receiver

Incidents such as these happen several times a day. We perceive an object that has caught our attention, and thanks to this object we understand something new, and this comprehension often has an impact on the rest of our experience, no matter how banal it might be. *There is a sign if it places something perceptual in a reciprocal presupposition with something that is cognitive.* We have a sign every time the perception of something stimulates in us an increase in knowledge, a sign that will often have a practical effect on what we are doing, desiring or looking for. Despite the philosophical tradition dating back thousands of years that, here in the West, has tended to separate the senses and the intellect, body and soul, the flesh and the mind, signs place these things in relation with one another.

Staying with the theme of traffic, if whilst I'm driving the flow of the traffic is interrupted due to a group of cars blocking the way (signifier), I will assume there has been an accident (signified), and this sign will have very precise consequences for me, for example, with regards to my timetable that day. Likewise, the thickening of clouds (signifier) will lead me to think of an imminent downpour (signified); the presence of smoke (signifier) indicates fire nearby (signified); a pawprint in the sand (signifier) suggests an animal has recently passed by (signified). It is not these phenomena themselves (clouds, smoke, pawprint) that are significant, but their perception by someone who, by empirically coming into contact with them, associates them with precise meanings (probable rain, the proximity of fire, an animal's trail). *The sign is* not, therefore, something that stands in for something else, as is often said, but *the relationship that a subject instals between two elements in which one involves the senses and the other, cognition.* It is a relationship that comes about thanks to the person who is there to perceive and think, and certainly not thanks to whoever created those phenomena. The animal that left its pawprints on the sand certainly did not mean to do so, for it to be seen, for example, by the hunter following it. Similarly, the pyromaniac who set the fire did not want in any way to be noticed because of the smoke fanning out around them. Nor did the clouds, an atmospheric agent, intentionally aim to herald the onset of rain. Just as (of this we can be certain) whoever was involved in the traffic accident had no intention of either causing the ensuing traffic jam, nor of that sudden build-up of cars announcing the accident to whoever was arriving in the area.

I.3 Communication and signification

Let's return to our island. There are a few more reasons to further consider what happened there. We have already stated one of them. Much like the other cases we have cited, the people that nonchalantly parked their cars on the side of the road in order to go down onto the beach had no intention of indicating their preference to passers-by. They would probably have preferred not to announce their presence, and spend the day on the beach without the crowds. It was us who, in search of a good place to swim, noticed that all those cars were parked there because there must have been a way of getting to the beach. In more technical terms, the sign is not produced by whoever makes the constitution of the signifier possible, but by whoever sees and interprets it. Of course, there are many signs that are intentionally emitted in order to communicate something, like when we point out an object to someone, when we give a lesson, when we ask the time, when we paint a portrait, declare love, write an article for a newspaper, broadcast an advertising spot, or send an email or text message. But there are many more, like those we have cited above, and which are probably in the majority, that are emitted in an unconscious way, without any hint that they are signs, and that only the receiver can perceive as such. Each of us is a carrier of hundreds of signs –the way we walk, the way we talk, how we dress, what we eat, the things we do or don't do – all of which contribute to the constitution of our social identity, the image others have of us. There are others – in our eyes, continuous issuers of signifiers which unintentionally offer to observers our interpretations, appreciations and judgments. Not to mention the places, natural or urban landscapes, or objects, buildings, other living things, even natural elements like the clouds in the sky. All of them are materials with the potential to become the signifying expression for spreading all kinds of contents, as long as there is someone somewhere who exists in the right conditions to carry out the act of interpretation. Contrary to what we generally believe, *the first motor of language*, of all human and social languages – or rather, of *signification* – *is not the sender but the receiver*. It is only with the hunter who catches sight of the pawprints, the old woman worried about the rain, the forest ranger who has to protect the trees, the driver who hates traffic, the holidaymaker in search of a beach, that a signified content that begins with a signifying expression can be generated. Traditional *communication* takes place when the transmission of a message is intentionally activated, when there is someone who wants to say something and so does anything they can so that their interlocutor acknowledges what they are saying. *Signification* is a more general phenomenon (that includes communication as a subset), in which meaning is understood by whoever manages to interpret a signified by perceiving a signifier – like we did on that day at the seaside. We must remember this difference between communication and signification because it will be fundamental.

I.4 Inferences and culture

Returning to our discussion, the question now becomes: on what basis is the receiver able to institute a relationship between expressions and contents? Is this relationship always valid and therefore easy for anyone to notice, or must there be precise conditions in order for this to happen? There is obviously nothing universal or necessary in this since it depends, in the first instance, on the mental mechanisms activated (or at least that are able to be activated) by the interpreter – mechanisms that have a different logical value depending on the case in hand. Let us take the example of the traffic accident recognised by the build-up of cars along the road. Is this a *deduction*, a reasoning that goes from the universal to the particular? It surely is not: there is no universal law that dictates every time there is a build-up of traffic along the road it means there has been an accident. Rather, this association usually takes place in the majority of cases to the point that, through *inductive* generalisation, an interpretative rule is installed retrospectively. A rule which, though not valid in all cases (there can be many other reasons for the traffic), is frequently correct.

Let us now return to our holiday fable: that case was not one of deduction or induction as not only was there no universal law to apply, there was also no previous experience that gave rise to a rule. Let's admit it: the meaning here was, to a large extent, the result of a guess. Those cars could have been there for all kinds of other reasons (there were no other car parks around, there was a party in one of those villas, the road ahead was blocked . . .) and we had never been to that island before. Our *inference*, the way of reflecting we used in order to connect signifier and signified, was therefore an *abduction*, a comprehensive law ("There must be access to the sea!") arrived at from a particular coincidence ("Look at all those cars!"), missing out the procedures of generalisation. The emotional dimension (that combination of enthusiasm, curiosity, carefreeness, expectation and joy that we were experiencing) contributed considerably to our construction of a new form of knowledge. Emotion and cognition worked in unison, once more casting aside the years of philosophy that wanted to keep them separate.

This abduction was not the result of similar previous experiences; but it worked, thanks to our *culture* of reference: that collection of knowledge, value systems, series of habits and ways of behaving in which we usually dwell, and that led us almost naturally to guess in that particular way rather than in another. On a small Mediterranean island these things happen: in order to spend a few peaceful hours in a beautiful place by the sea, people tend to abandon their cars at random. In Sweden, for example, that sort of thing would never happen, and a Swedish person would never reach the same conclusions when faced with a similar situation, no matter how rare that might be: "Why are there lots of cars at the side of the road?" (signifier), "There's going to be an earthquake!" (signified). The exceptional nature of the event in that part of the world can only lead to catastrophic conclusions.

So, we can say then that the cognitive inferences (with varying degrees of logical stringency) that we use in our own interpretations, associating expression and content in our everyday experience, are in no way personal, subjective, idiosyncratic. They are based on precise *codes*, a number of formal systems that regulate the possible associations between signifying expressions and signified content. These codes transcend the choices of the individual and often impose upon them the use of fixed mental categories that lead to locally pertinent interpretative conclusions. Moreover, these codes are in no way universal or necessary: they are not objective; but this does not necessarily make them subjective. They are established clusters, with varying durations, of collective ways of thinking and acting, of desiring and expressing preference. They are dictated and maintained by social pressure, which also modifies them when it holds them to be outdated (just think of dress codes, which are much less constant than those related to food).

The codes are social and cultural customs, interpretative habits that take on the appearance of a law (without actually being one), and no individual inference is possible without these as its starting point. If we and the many other people who had left their vehicles on the side of the road that day had not shared the same culture, if we hadn't belonged to the same society, that sign (unruly collection of cars → "path to the sea") would not have existed. In Sweden, as in Germany, Denmark or Great Britain, in half of the rest of the world, codes that lead to this sort of conclusion simply do not subsist. There will be unruly collections of cars and no lack of paths to the sea, but they will not give rise to the shared rules that connect (the perception of) the car and (the discovery of) paths. Let us say it once more: the sign is not the thing nor the idea, but the form taken by the relationship between them.

I.5 Difference and values

In others words, codes are anthropological conventions. This means that the possible (and impossible) forms of association between expression and content that they establish, most of the time are not bound to stringent rules, to logical forms, and perhaps not even to reasons of economy or convenience. Rather, they depend on *values*, orders of preferences, tastes and revulsions, of extreme complexity which, most of the time, are as arbitrary (if viewed from the outside) as they are indisputable (if viewed from the inside). As such, the different interpretive codes of the Mediterranean island and the Swedish coast are bound to different cultures, which are based on different value systems – value systems that could be, in the first case, that of the individual, of spontaneity and carefreeness, of a certain customary flexibility; and, in the second case, that of a regulated social-mindedness, of a shared, ordered and undoubtedly more rigid emancipation. "People of love" versus "people of freedom"? That would oversimplify things somewhat but it gives us the right idea.

The notion of value is composite and takes on many forms. We will return to it many times over, not least because it is very close (if not equivalent to) another notion we still have not defined: that of *meaning* – which lies at the heart of every semiotic reflection, much more so than the concept of sign. We are using the notion of value in the negative, in order to scale down the importance of the logical and cognitive aspects of signification when discussing semiotics, as well as the economic, materialist, and essentially rationalist explanations of societies and cultures that tend to be dominant. If we consider it in the positive, the idea of value also serves to demonstrate how social codes, despite transcending individual choices, are in any case bound to the ways in which the individual *assumes* them, making them their own, mediating between collective values and individual ones.

Let us go back to our little island and reconsider the situation. In macroscopic terms, we can say that *the sign is born out of difference*, through the perception of a visual *discontinuity* (that could also be acoustic, olfactory or linked to any of the senses in general). This discontinuity was between a landscape where everything ends up being the same, a road where the various elements disappear from our attention – a place that is progressively insignificant – and the sudden apparition of the parked cars that are noticeable by their difference (the glare of the sun on metal, the industrial colours of the car bodies, the unexpected heat they give off). The perceptive residue causes something meaningful to re-emerge, something significant, something of importance because it is positively valorised by us when compared to the insignificance of everything else. The road signs, the signs for shops and B&Bs, held very little value for us because they had no relevance to our plans for a day at the seaside. Actually, they distracted us from our plans, suggesting alternative, if not clashing, itineraries. On the one hand, therefore, we have a general euphoria given by the state of being on holiday that allowed us to be relaxed about ambling along without any sense of timing, with no commitments or obligations of any kind. On the other, there is a kind of itinerary of knowledge involving an environment that is unfamiliar to us and that forces us to always maintain a degree of attention, to observe and weigh up, to connect, compare, judge, *value* things, objects, atmospheric agents, the folds of the landscape, the alternative routes along the road . . .

This is the double status of values. On the one hand, they reside in what we were searching for, in making the desired object of our mini-adventure significant, important. In this case, the value is that which is being aimed for, it directs our actions and passions, giving a very precise meaning to each of these within a sort of implicit narration. On the other hand, the value arises through continual comparison with other objects and other values, in the difference between the elements and the forms of their relationship. Shall we stop at the newspaper stand? No, let's keep going to the beach. Shall we buy sunscreen? We don't need any. Shall we visit the little town on the top of the hill? No, let's continue our search. Shall we eat a sandwich? There's no time. In our eyes, the value of the beach was constituted out of difference, and somehow it slowly grew when faced with what we could potentially do instead of going on to find it, but that we ultimately decided against. The social code that had already been set in motion, allowing us to associate the perception of the parked cars (in discontinuity with the previous landscape) with the idea of a path down to the sea, worked in the background of this short *story*. Outside of this story, no one would have taken that code, made it their own, personalised it, and used it.

I.6 Relations and narrations

We are beginning to understand the extent to which the sign we are discussing owes its existence to the many constitutive mechanisms that are hidden and, for this reason, fundamental. The sign is only the tip of an enormous semiotic iceberg: it is what appears immediately and very clearly (like the cars that mark the path); but that works only because, on the one hand, it breaks down into smaller elements that are articulated with one another, and, on the other, because it contributes to the creation of larger entities through relationships with other signs. This is something that is true for both the signifying expression and the signified content: both are at the same time composite elements and pieces of larger elements. The cars signify through their incongruous assembly in a place where we would never expect to see so many; but, also, through the way in which they are badly parked at the side of a road with no sidewalk, and through the way they grow more numerous the closer they get to the path and vice versa. Similarly, their signifier is composite: they indicate the presence of a path but with no certainty. Something else might have attracted so many people, something similar like a jetty or a lido with sun loungers or a panoramic view. All of this is then inserted into the environmental framework, be it geographical, urban, administrative and even political structures of the island. The collection of cars experiences different relationships with the identical hedges that block the view onto the villas, the shrubs of bougainvillea or jasmine that add colour, with the road signs and those advertising other things. The small path that leads down to the sea contrasts with the tarmacked road we are currently on, representing an invisible deviation, and all the other roads that branch out from our own. All of this is inscribed in a story: the one we have lived and which personalises the situation, reconstructing it from a precise perspective of values, and culture. Two Swedish people on holiday on the same island would have never enjoyed that crystal clear water, because it did not appear in the guidebooks. Not only would they have been missing the rules of the code that combines the signifier 'car' with the signified 'path', but they would miss the entire dynamic *text*, the overall network of relationships within which that sign is formed.

Signs function because they interweave within texts, just like words which are pieces of language that function if and only if they are used within phrases and discourses that, by placing them in a condition that makes them signify, transcend those very discourses and phrases. Words, like all signs, are also the variable result of constant interactions between smaller elements (morphemes, phonemes, sound traits) and are at the same time entities that contribute to the composition of larger structures (phrases, texts, discourses). From our perspective, texts are not therefore just books, written documents, linguistic 'established' materials as it were, but are *any portion of the world that, through the possession of fixed boundaries and a precise internal articulation, makes itself the carrier of any configuration of meaning* – as we see in our anecdote about the trip to the island which, albeit banal, has all the internal and external characteristics of a text.

Semiotics does not deal therefore with signs, but with *signification*, which in order to function – to be produced, to circulate, to be received and finally transformed – must make reference to texts, the units of meaning that various societies, historical eras and cultures use in order to communicate their own basic values. Anything can, in principle, be a text. Any material present in the world – physical and biological, physiological and corporeal, geographical and social, cultural and historic – can be used (if correctly organised) as a signifying expression used as a vehicle for certain content. We will see this over the course of this book, and we will also see how everything that is human and social – from food to clothes, from how we spend our time to social hierarchies, from the structure of the cosmos to our urban fabric, from political organisation to religious rituals and the myths that tell of them – can become the signified content communicated through any possible substance of the world. Text should not be understood (as we have seen) as a disordered collection of signs, any amalgam of things that stand in for others, but as a *fabric of relationships*, a web of forms that bind substances and aim to produce, set in motion and justify any unit of meaning – be it just a trip to the sea.

I.7 Form and substance

Our further reason for using such an ordinary little story as a way of introducing this book on the semiotics of text will now be clear. The texts semiotics deals with, before they are gradually established and valorised by any culture as such (books, but also images, audiovisual texts, plays and musicals, today's Facebook posts, Tweets and YouTube videos), are those that we encounter in our everyday lives, the *profound and almost always involuntary form of our human experience, both social and individual*. The fact that the everyday, the ordinary, the banal are shared experiences, does not mean that they are elementary, immediate phenomena. Most importantly, they are not so much evident as always apparent. The everyday is full of signification, meaning insinuates its way into everything, especially into that which is most obvious: our bodies, food, architecture, clothing and so on. We could say that signification is the amniotic fluid in which the human being, the social subject, groups and collectives exist, and that there are many more things that hold meaning for us than there are things that do not. The evidence of everyday experience is founded on deep structures that are hidden, unaware, unconscious. They are semiotic and therefore textual.

It is the task of semiotics to reconstruct these structures, to show them and clarify the mechanisms with which they function. Often, this is a movement against common sense, which instead tends to hide them, deny their existence and be annoyed when they reappear. If, as a semiologist, I attempt to explain the signifying value of a dish, and how by cooking or eating it we are spreading its cultural and social meaning, the response is that we are only cooking or eating the dish because it tastes good, or because it is good for us, or because it is the only thing available, thus entirely denying its semiotic value. The same thing happens with clothes or interior design, with exercise at the gym or shopping at the supermarket: all phenomena filled with signification that are, however, cloaked with utilitarian, economic, rational, ethical rationales for their existence. Meaning entrenches itself behind motivations of different natures that tend to suppress, or at least hide, it.

When using our language, an extremely complex entity, we bring into play a knowledge that we are unable to explain. We all adopt very sophisticated rules on phonetics, grammar, morphology, syntax, etc., but this does not mean that we would be in any way capable of explaining them if someone were to ask us to teach them. Actually, unless we are accomplished linguists, we have absolutely no idea what it is all about. Language is a practical knowledge and we do not therefore think about it. And when we do, we almost always end up making mistakes or errors, like insisting (wrongly) that the smallest units of language are its words. Then scholars of the subject intervene and clarify, not without clashing with the resistance of the speakers (who ferociously defend their own naïve positions), that words play a secondary role, and that other elements are truly important in language, such as the objects of phonology, morphology, syntax, and so on.

In the same way, most of our experience – no matter how casual, spontaneous, immediate or banal it may seem – is based on a network of highly complex significations, on codes and texts that, even though we may use them, we are not necessarily capable of explaining. Is everyone who goes on holiday to the Mediterranean islands aware of how many semiotic mechanisms will be brought into play when they want to find access to the sea? Obviously not, they are simply there on holiday. They cannot even imagine how much implicit semiotics they are using, nor do they want to know. Similarly, when we dress ourselves, eat, go to the delicatessen round the corner, live in an apartment, meet with friends and relatives, talk about everything and nothing, make love, play sport, go out to dinner, drive a car, walk through the city centre, travel, not to mention when we consume music or film, television programmes or newspapers or surf the internet, when we go to the mall, we are constantly sending and receiving texts of varying size and importance, with very different signifiers, which most likely function using the same mechanisms, the same *forms of expression* and *forms of content*.

We will learn, for example, to understand the way in which these texts that weave the fabric of our daily lives tend to have a profoundly narrative form. This is not evident most of the time, but it is nevertheless present. From this perspective, storytelling is neither a new communication technique nor a recent discovery. As traditional cultures were well aware, and as the media culture that builds on those traditional cultures knows equally well, stories are the lifeblood with which we give meaning to our actions and passions. The events which happen have an accurate value for us depending on the specific role they play within narrative structures, whether individual or collective. Myths, legends, narrative folklore of all kinds, as well as the media narration of our own era (journalistic, televisual, cinematographic, scientific, advertising, and so on) are all networks of meaning. It is no coincidence that in this introduction to the semiotics of the text we chose to begin by telling a story, a silly little tale that, despite being banal, revealed itself to be rich in signification. The Great Narrations of a time are undoubtedly going through a moment of crisis, despite often reappearing where we least expect them, as is the case with religious fanaticism, migration, and even certain media sagas. The little stories, however, remain and give meaning to our lives, and life to our language, to our human desire for communication and need for reciprocity.

Recommended bibliography

On the history of the semiotic structural paradigm, see Bertrand, Bordron, Darrault & Fontanille (eds.) 2019; Broden & Welsh (eds.) 2017a, 2017b; Jakobson (1974) and Eco (2002); see also Manetti 1996 (on ancient semiotics), Marmo 1997 (on medieval semiotics), and Hénault 1992 (on contemporary semiotics).

Handbooks, dictionaries, and introductions to structural and textual semiotics include: Bertrand 2000; Entrevernes 1979; Fabbri & Marrone (eds.) 2000, 2001; Fabbri e Mangano (eds.) 2012; Hébert 2018, 2020; Hénault 1979; Hénault (ed.) 2002; Klinkenberg 1996; Browen & Ringham 2000; Pozzato 2001; Traini 2006; Volli 2002; Zunzunegui 2005.

The theoretical basis of the semiotics of the text may be found in Greimas 1983a, 1983b, 1987a, 1987c, 1988, 1990, 2017; Greimas & Courtés 1982, 1986.

On Greimas' theory and Paris School semiotics, see: Arrivé & Coquet (eds.) 1987; Beividas 2017; Bertrand, Bordron, Darrault & Fontanille (eds.) 2019; Broden & Walsh-Matthews (eds.) 2017a, 2017b; Coquet (ed.) 1982; Darrault-Harris (ed.) 2016; Kersyté (ed.). 2009; Landowski (ed.) 1993, 1997, 2017; Parret & Ruprecht (eds.) 1985; Perron & Danesi 1993; Perron & Collins (eds.) 1989 (the most important reference of structural semiotics in English, with articles of the principal scholars); Rector 1979.

For the development of structural semiotics of the text, see: Ablai & Ducard (eds.) 2010; Akrich & Latour 1992; Berthelot-Guiet & Boutaud (eds.) 2014; Biglari (ed.) 2014; Coquet 1997, 2007; Courtés 1976, 1995; Fabbri 2007; Fontanille & Zilberberg 1998; Geninasca 1997; Gimate-Welsh (ed.) 2000; Hénault (ed.) 2020; Marsciani 2012a, 2012b; Quéré 1992; Parret 2018; Rastier 1987, 2001; Violi 2001; Zilberberg 1982, 2006.

Specifically on the notion of text (from a linguistic, literary, and semiotic point of view): Adam 1990, 1992, 2000; Barros 1990; Barthes 1977; Charles 1995; De Beaugrande & Dressler 1981; Frölicher, Günther & Thürlemann (eds.) 1990; Perron 2003; Petöfi (ed.) 1983; Pozzato 2001; Rastier 2001; Van Diik 1972, 1977; Van Dijk (ed.) 1985; Wahl 1972.

The principal journals with articles on the semiotics of the text: *Actes sémiotiques, Degrés, deSignis, E/C, Il Verri, Lexia, Ocula, Topicos del seminario, Semiotica, Signata, Sign Systems Studies, Social Semiotics, Versus.*

II Basic principles

II.1 Historical emergences, plural filiations

Textual semiotics was born in the twentieth century, bringing together a whole host of themes and numerous areas of research. It concerns all the human sciences that have based their own existence on a number of fundamental concepts that textual semiotics, by re-proposing them, has made its own. These include: the importance of language and communication in the social sphere; the structural precept of the primacy of relationships over elements (correlated to the idea that an entity only acquires value if it is placed in a direct relationship with other entities from the same system); the principle of pertinence, according to which relationships between elements can vary according to the perspective from which they are observed; the hypothesis of a profound layer – or immanence – found in every empirical manifestation of socio-cultural phenomena (that explains its underlying logic, the rules of combination, and the ways of combining a small amount of invariants and highly variable elements). If social sciences can boast about their own epistemological rigour, it is because they do not search out mathematical logic or principles with their foundations in physics, but because they study the way in which human and social meaning is produced, articulated, made manifest, and the way in which meaning changes: in short, the social sciences search cultural texts.

The path for achieving this epistemological autonomy – and consequent critical awareness – has been a long one. On the one hand, there have been slow and patient studies in the field aimed at locating interesting material with which to construct theory-laden case studies: in folklore, anthropology, sociology, linguistics, literature, psychoanalysis, mythology, narratology, mass-media studies, design and so forth. On the other, there is the need for continual comparisons between these worlds that remain distinct, in an attempt to prove shared dimensions and relative specificities, general forms and particular substances. Semiotics has made a significant contribution to this two-pronged approach, providing a methodology for the human sciences and a general theory for the ways in which meaning, through articulation, can achieve the right conditions in order to signify.

The route to epistemological autonomy first appeared within linguistics, linked to the great theoretical proposals of Structuralist masters such as Saussure, Trubetzkoy, Jakobson, Hjelmslev and Benveniste, who wanted, and knew how, to rearticulate the traditional problems of philology and comparativism. In a similar way, the works proposed by formalists (Shklovsky, Tomaševskij, Tynjanov and Jakobson again) in the field of literary studies took up and reinforced traditional rhetoric, poetics and stylistics. The research programme followed by linguistics and literary theory (which benefited both) led to the identification of that shared research terrain that is textuality, which went beyond phrasal syntax (for linguistics) and aestheticizing impressionism (for literary theory). Something similar happened in the neighbouring fields of folklore and mythology, where scholars such as Propp and Lévi-Strauss introduced formal and structural analysis to the research of popular and ethnic narration, reconstructing the possibility of locating variable elements shared by very broad anthropological traditions beneath the extraordinary variety of local and individual variants. Somewhat slower, though no less effective, was the emergence of textuality in the study of mass culture, in which authors such as Barthes, Eco, Fabbri and many others were able to demonstrate the complexity of the messages emitted by mass media, on the one hand, and the possibility of identifying shared underlying and ideological textual structures in seemingly fleeting products, on the other.

From all of this came, in the 1960s, narratology, i.e. the study of the general laws of narration, regardless of aesthetic significance, social function or means of communication. By utilising the results of research carried out in different fields (literature, folklore, mass media, mythology, everyday jokes, etc.), Barthes, Greimas, Eco, Todorov, Genette and many others put together a grammar of narration, similar to, but not interchangeable with, that of linguistics, and founded on textual principles (Barthes ed. 1966). Rather than limiting themselves to studying literary *works* with a narrative character bound to an aesthetic valorisation and a predetermined cultural role, narratologists considered narrative *texts*, i.e. all the possible communicative products constructed in accordance with the deep-rooted laws of narration. Initially, literary texts continued to be the main focus, quickly becoming the object of intense and thorough textual analysis. Just think of books such as *S/Z* by Barthes (1975b) or *Maupassant* by Greimas (1988), entire volumes of analysis devoted to a single brief novella that paved the way for hundreds more works. But, just a short time after, narratological research was extended to the study of journalism, cinema, television, comics, advertising, painting and photography, architecture and design, as well as delving further into the well-established fields of ethnic cultures and mass communication.

And so, bit by bit, the text becomes a semiotic category in its own right. The science of signification further broadened the notion of text, using it to study highly diverse cultural manifestations. In this way, television schedules, advertising campaigns, informational flows, communication platforms, verbal conversations, online interactions, cooking recipes, marketing strategies, subway stations, buildings, even whole cities can be examined as texts *from a methodological perspective* even if they are not seen as such *from an empirical perspective*. It is conceivable to find the same formal properties in these phenomena as we do in traditional texts. The text as understood from a semiotic perspective is no longer a thing, an empirical object, but a *theoretical model used as a descriptive tool* under certain specific and explicit epistemological conditions. The text allows us to reconstruct the profound formal dispositives of any object of knowledge within the science of signification. It is the formal model for explaining all human, social, cultural and historical phenomena.

From this comes the birth and development of socio-semiotics. Following a seminal paper given by Fabbri (2018 [1973]), who offered the semiotic gaze as a methodological antidote to the theoretical 'evil eye' of sociology, many authors such as Floch (2000, 2001) and Landowski (1989, 1987) began to explore the possibility of a formal and semiotic study of social facts such as advertising, politics and journalism, fashion and design, food and daily lives, objects and sensory experience, paying greater attention to their broad social and cultural significance rather than their textual structures. By analysing not already established works but much less tangible phenomena such as situations, conjunctures, practices, uses, sensory and bodily experiences, communication and media flows, the interactive nature of new media and so on, socio-semiotics lays bare the dichotomy between 'text' and that which surrounds it: 'context'. The so-called context, if assumed as part of a coherent project of description, has its own semiotic consistency as a text. Where the linguistic perspective (even in its pragmatic declination) distinguished between linguistic phenomena and extra-linguistic phenomena, for socio-semiotics, this difference can never exist *a priori* because anything can be signifying and social, communicative and factual, textual and experiential. From a socio-semiotic perspective, context is (put very simply) that which is not pertinent to textual analysis – a pertinence that is determined by the social culture long before any analysis takes place. And the text is not a material excuse for possible interpretations that integrate the context or that even justify its existence, but the formal dispositive through which meaning articulates and manifests itself, circulating in society and in culture. This is, therefore, *the specific object of study for the semiologist*, who, through analysis, must attempt to reconstruct its forms and dynamics, its internal articulations and levels of pertinence. The text is not a given, a phenomenal fact, but the result of a double construction: first, the socio-cultural configuration, and secondly, the analytic reconfiguration. From this point of view, the text must be the result of negotiation between cultural dynamics in a continuous interweaving with other texts, other discourses, other languages. There is nothing closed about it. Rather, it is open, has permeable edges, and is always ready to adapt to other textual configurations, to translate itself into other languages, in that inter-textual, inter-discursive, inter-media chain without end that is the semiosphere.

II.2 How we recognise textuality

Though it is bound to a cultural dimension that can change some properties, textuality can be recognised when certain basic characteristics are present. We will attempt to quickly run through them here.

a) First and foremost, the principle of *negotiation* is fundamental as it immediately eliminates every form of ontology. There are no texts with precise expressive substances or preferred configurations. A sign, an emblem, a logo can all be, under the right conditions, texts, just as, under other conditions, they can become elements of broader textual occurrences, such as an entire brand strategy. The sign and its smallest elements change their role each time. A word can be the carrier of an inherent signified (and become a text), but at the same time it is made up of monemes and phonemes, and it is a single entity within a phrase that, in turn, is a tiny element of a whole speech. Similarly, the relationship between the text and its parts, or between the text and the macro-text, is variable, and it is determined by its assumed pertinence to the analysis each time. In the text, everything is negotiated, starting from its borders, be they spatial or temporal, physical or semantic. As such, a television opening sequence, the frame of a picture (Stoichita 1989), the cover of a book (Genette 1997b) or the fourth wall at the theatre are entities that, no matter how habitually we consider them obvious, are the result of negotiation. And they can in any moment return to being the object of negotiation (actors who choose to ignore the fourth wall and descend into the audience or the scholar who decides to take not a single work but an entire genre as their field of research).

b) The idea of negotiation is constitutive for textuality because the fundamental characteristic of the text (and of languages and semiosis in general) is that of *biplanarity*: i.e. the reciprocal assumption of two planes – expression and content – each equipped with a purport (relatively non-pertinent) and a form (that is, instead, constitutive). A text is founded by the basic solidarity between a *form of expression* and a *form of content*, which leads to the emergence of signification and gives it substance. In other words, what is important for textual signification is not the choice of the substances used (audio, visual, etc.) nor the choice of the subjects to be communicated, but the relationship between the two, which can exist only if both things – substances and subjects – are articulated, manipulated, formed. So, there is no particular audio purport but a precise expressive configuration; and there is not a general subject, but a specific way of discussing it, of tackling it, of ordering it and forming it. It is only when the two operations of formation have happened (at the same time) that human and social meaning is given. Thus, we understand once more the reason why the text is not an objective fact but a formal, dynamic and ever-changing construct. The text is the process of two simultaneous operations of formation being placed in relationship with one another; operations that start out as arbitrary but are necessary and in continual transformation.

c) Our third point arises from the principles of negotiation and biplanarity, that of textual *closure*, which is never definitive except within usual communicative canons that can be more or less codified. For example, a ceremony has clear limits: precise performative statements dictate the opening and closing of the event ("*Introibo ad altare dei*", "Go in peace", "I declare this session open – or closed", and so on) (Austin 1962). A social gathering, such as a party in the town square, has less rigid boundaries unless the authorities intervene to regulate timings. Generally, the textual boundary, semiotically speaking, is always variable, meaning that even the circumstances of the text's production can become part of it. Just think of those so-called interactive, cross-media texts in which the receiver's response contributes to the author's construct of the text. There are also all those mass media products in which the presence of a particular director or actor in a film predetermines the plot, setting in motion a series of expectations about its outcomes that can be met or otherwise. From this, we can deduct that textual boundaries are not ontological but must, nevertheless, exist. They are necessary in order to highlight the constitutive discontinuity, that basic perception of difference without which no signification could take place or even be suggested.

d) Even more than closure, it is therefore the *consistency* of the text that is important, with this term understood in the sense of the famous Saussurean structuralist admonition that in a language *tout se tient*. It is the consistency of the textual whole that generates both its internal articulation (its structure) and its edges (which can exist only as a negative, as a progressive dispersal of the elements, an interminable drip from a tank that nevertheless holds and holds back). The notions of formal coherence and semantic cohesion as discussed by textual linguistics can be traced back to this principle of structural organisation, thus avoiding a certain tendency to *a priori* regulation that they can convey.

e) In this sense, the textual consistency strongly demands internal transformation, intrinsic *processing*. Indeed, beyond its systematic (i.e. paradigmatic) organisation, a text also has its own syntagmatic evolution that can manifest itself as temporality (for example, in linguistic or audio-visual communication), spatial organization (like the visual elements in an image), or narration (in which precise rules dictate that what it is at the beginning and what it at the end are never the same thing). Despite the circularity of certain formulaic folklore genres, the perception of the elements, and therefore their semantic value, is always modified over the course of the text. In the text there is a profound narrative organisation in which a pragmatic/passionate programme and a clash between subjects lead to a subjective transformation, be it individual or collective.

f) There is another point to remember here, that of the *multiple levels* within the text, i.e. the fact that the comprehensive semantic formation of a textual whole can be understood to varying degrees of simplicity or complexity, in a more abstract or literal way. In Greimas, this principle is called the *generative path of meaning*, which we will look at later on (Greimas & Courtés 1979). In general, however, both in that moment of lived experience and that of textual analysis, the same meaning can be expanded to varying degrees over various numerous elements or condensed down into a few defining traits. Within its structure, every text is articulated in multiple levels of pertinence. The deepest is a common background and an interpretative grid for every social and human signification, whilst in the most superficial level an enunciating subject (enunciator or enunciatee) positions the text within a social communication that allows it to circulate.

g) This is linked to the idea that every text also contains within it, beyond its enunciated content, the image of its communication (or *enunciation*): the principles for its own function, the criteria for its production and use. In other words, its instructions for use. Beyond its real sender and receiver, within the text we find their images as simulacra, the enunciator and the enunciatee, abstract *actants* that can be transformed in various ways into *actors*, and that can be much more performative than a sender or receiver of flesh and blood. For instance, a brand is a company's communicative image that corresponds more or less to its production and commercial reality. However, in the marketplace it ends up being much more performative than the company itself, affecting real consumption processes. Similarly, the target is the image of the consumer, but the brand ends up generating them, instructing them, influencing how they consume and even their lifestyles. It is the image of the public found in television broadcasts that determines the receptive behaviour of the audience. It is the idea of the interlocutor in my discourse that ends up generating it. It is the planned behaviour within a building or apartment that leads the person living in it to behave in a particular way, and so on, including the possibility of inverting positions so that, as De Certeau (1984) says, it is the praxis of consumption that constitutes the texts' meaning: the tactics for use can upturn the structures that have brought certain messages into being. In one way or another, the level of enunciation lies within the text but is always turned towards the outside, to the cultural world that, in one way or another, will re-say it in its other texts.

h) This brings us to our last point, that regarding *intertextuality* and *translation*. If the inside and outside of the text are reciprocally defined, the relationship between one text and another is constitutive for both of their identities. Textual closure is not the isolation of a monad that does not communicate with other texts. Every text is, according to Barthes (1975b), a perspective of citations because it contains within it that which Fabbri (2017) refers to as 'invitations' to read further texts. Every writer creates their predecessors, Borges said. In other terms, intertextuality is not only the philological return to sources, nor the hermeneutic history of the effects of reading, nor even the postmodern wink at past literary authors and traditions. If anything, it refers to the pre-existence of other texts within any text (what Eco 2002 calls *encyclopaedia*), or the presence of a discursive basis in a text that inscribes it within a culture through a series of cross-references that place the process of translation (which is essentially the same thing) at the centre of its structure: inter-linguistic or intra-linguistic, inter-textual or intra-textual, inter-discursive or intra-discursive. Meaning is given through transformation, and texts are the ever partial and momentary results of operations of translation.

II.3 On the principle of pertinence

"It depends" is the most frequently heard refrain in semiotic textual analysis – that is a descriptive and interpretive instance that does not study objects, things, or given realities, but relationships, structures, systems. And thus doing, it delves into the way in which these relationships are constructed. For textual semiotics, there are not isolated signs, elements that signify on their own, but only networks of signs that, by entering into relationships with one another in many different ways, produce processes of signification and texts. Holding it all up is the principle of pertinence, according to which every single element *depends* on the others with which it interacts, and only their interaction leads to the construction of textuality.

Let us take a very banal example: the first letter of our alphabet, A. What does it mean? Everything and nothing, as its meaning changes *depending* on the system in which it is placed, the text to which it belongs. It is the first letter of the alphabet, if we take the system of the alphabet as the point of view from which to consider it, but it is also the suffix designating the female singular in most names in the Italian language. If we change perspective again, it denotes the third person singular in the present tense of the verb 'to have' in French, or the indefinite article in English. It is the first element in the vowel system, a simple preposition in French, but it is also the symbol for Anarchy, or the code for Austria, or the initial of someone's first name. And so on. In short: it *depends* on the context into which it is inserted, or rather, it depends on the text that proposes it as an element within a network of relationships. This is pertinence: the choice of perspective which reveals the value of an element in relation to others within a system.

Let us use another example: the flight of a flock of birds. We know that the ancient Greeks interpreted this as an ominous sign, but it required an augur, in accordance with a set of rules, to correlate the shape of the formation in the sky to the result of the next battle to be fought, to give their opinion to an army ready to burn for the *polis*. An ornithologist today would try to correlate that same flock of birds to other formations made by the same group of birds, or to that of other kinds of birds, perhaps in different skies. But, for someone studying the climate, things would be different, as they would be for a landscape artist, a tourist photographer or a lover trying to seduce the object of their affection. Once more the element is the same, but it is the point of view that changes, and with it, the element's meaning. The same thing happens with words in a language. Words don't mean anything outside of the phrases in which they are found; or, rather, they mean too many different things without those phrases. Proof of this can be found in the dictionary when under a term I discover a list of its possible meanings with relevant examples. It is the comprehensive meaning of the whole that constructs meaning in each term, not the opposite. Not to mention natural symbols such as colours, plants, animals and so on. We generally think that colours have a symbolic signified (green = hope, red = passion, yellow = jealousy) but this signified is in no way stable. It changes according to the systems that regulate the numerous possible relationships between chromatic tones (in traffic lights green is not hope but an invitation to cross the road, just as red is not passion but a prohibition to cross that same street). Take the triangle: does it have meaning? We might say it has too many: it is a geometric entity, but it is also a symbol of God according to certain Christian beliefs, it is a warning sign on roads. The triangle also makes us think of an open relationship, the masons, even the occult. One last example is characters in a story. What do they mean? Once again it is not enough to point out what kind of person they are (good, bad . . .) as their value within the story depends on the other characters with whom they form relationships, on what they do, what they think and what they feel, the narrative roles they play and so on.

We will return to this – actually we probably won't talk about anything else – as textual analysis is a procedure that is as rigorous as possible in tracking down the bonds of pertinence that constitute a particular configuration of meaning, the series of different gazes that must be assumed in order to reconstruct the signification of a certain signifying core. In other words, textual analysis consists of the hierarchy of questions that the text must be asked, questions that would otherwise be opaque and resistant. The *explanation* of its forms (analysis) leads to its fullest *comprehension*, and a renewed interpretative capacity towards the text. We must avoid the mistake of starting with single elements (be they words, signs, objects, graphic marks, figures, characters or anything else) and then going in search of the conditions that make them signify. The textual perspective works in the opposite way. It directly considers the texts, the interweaving of elements, without any need to isolate from the network they are in, relationships. So, when we find ourselves before a text, faced with the problem of reconstructing its articulations, we must examine the system of its pertinences, the levels of meaning in which it is structured.

II.4 The generative stratagem

What is meant by 'level of meaning'? In order to understand this, it is necessary to introduce an elementary principle that is present in all languages: that of the expansion and concentration of meaning. Meaning can, in fact, be expressed in ways that are more or less concentrated (for example, in a reduced collection of signs) or more or less expansive (as with an entire discourse that explains meaning in its tiniest details). Take a story, for instance: I could sum a story up in a single phrase ("Ulysses returns to Ithaca"), or I could talk at length about it in the form of an entire epic poem (and write *The Odyssey*). I could utter a sequence of words "a man, a woman, an apple, a drama", or I can read in the book of *Genesis* or in one of its numerous interpretations the meaning of original sin. I can dictate a few lines in a journalistic note or I can compose a long article in eight columns. I can sum up the basic principles of a brand in its logo, or rely on complex advertising campaigns to communicate the same thing to consumers. If we look closely, this is a principle of *paraphrasing*: any semiotic object can be paraphrased, expanded or concentrated depending on the case, depending on specific needs, on the communicative objectives and on the strategies within which it is inserted. The levels of meaning within a text are nothing more than the various possible ways in which it can be paraphrased, from the most simple and abstract way to one that is more complex. Analogously, in Borges' famous paradoxes there is the person who imagined a map of the Chinese Empire on a 1:1 scale, perfect but useless, and who rewrote *Don Quijote* exactly as Cervantes had, believing he had written an entirely new book.

The only way of saying something with a certain signified is to translate it into another signifier, changing it a little. Passing from one level of meaning to another, the signified never stays exactly the same. It loses or gains something depending on the case. But this is the game of analysis: repeating, on another level, the meaning of a text, collecting invariable and variable elements, shared cores and superficial differences. In doing so, what is missed on one level becomes clearer on another, until the analysis ends when all described objectives have been reached. This is why we talk about *generativity*, or in a more technical way, the *generative path*. During analysis, there is the need to reconstruct an underlying level of meaning that sheds light on that which, on a more immediate level, cannot be understood. We read tales, look at images, listen to songs, watch a film, walk along the road, use an object, eat something, and it all seems obvious, perfectly understandable. However, we only have to stop and think for a moment about these experiences and we can suddenly no longer understand their meaning, the process of their generation, or describe how they work. The aim of textual analysis is therefore to reconstruct the hierarchy of underlying layers in these experiences, the organised series of questions that must be asked in order to examine their internal articulation.

In doing so, analysis replicates in a rigorous and controlled (i. e. scientific) way the same gesture that various societies, cultures, historical eras make when they constitute, disseminate and transform their texts, whatever form they take, whatever name they may be given, no matter their level of self-awareness. Take the example of language. We conjugate verbs, we decline lists of complements, we navigate complex paradigms with extraordinary nonchalance, we use highly complicated rules that linguists struggle to reconstruct, and yet we don't know we are doing all this when we speak. We might have a vague idea, but it is not a given that our idea is correct. For example: we believe we are putting words in order when we are actually building phrases. Linguistic syntax is one of the underlying levels that dictates the rules for the composition of phrases, just as narrative grammar (as we will see) is the profound level that allows us to explain a large number of social and cultural discourses.

In text semiotics, as we have mentioned, this idea of textual analysis as a passage from one level of pertinence to another, as a means of simulating the levels' generation, has led to the proposal of a very general theoretical model known as the *generative path of meaning*. This is a theoretical tool that is particularly effective since it allows us to distribute and articulate the various observations it is possible to make when faced with any one text, hypothesising the possible stages of its progressive constitution. According to this theoretical model, the meaning of any text is articulated through signification according to levels of pertinence placed at various depths, in order of complexity and tangibility. As such, the most profound levels are abstract and simple, whilst the superficial ones are instead more tangible and complex.

The profound level of the path is that of *narrative structures*. The basic articulations of any universe of meaning, and therefore any text, use a narrative logic, so that narration (understood as the guided sequence of actions and events) is a *profound form of any human and social experience*. Narrative structures are subdivided into two layers. The first one, fundamental, is the 'semiotic square', where signification occurs through simple relationships of opposition, contradiction and complementarity, which constitute the values at play in the various semantic micro-universes. Signification is then enriched – at the second layer, anthropomorphic – by narrative programmes, in which actants and modalities interact. This narrative level of the path takes into consideration the *semiotic invariants*, those phenomena that can be seen in an almost identical way in all texts despite those texts apparently being very diverse. This is why the level of narrative structure is assumed to be general, valid in almost all cultures.

In order to produce semiotic differentiation and variety, it is necessary to move to the level of *discursive structures*. Here the relationships, the values, the actants and the modalities are enriched by both actors, spaces, times (syntactic components) and themes and figures (semantic components). The involvement of narrative structures in the discourse (known as *enunciation*) leads to the production of *semiotic variations*, or rather of the various kinds of discourse (literary, advertising, journalistic, philosophical and so on): every discourse is a different way of enunciating the narrative structures, of adapting profound semiotic constrictions.

It is only thanks to the mechanism of *textualization*, however, that the various discourses receive those substances of expression that allow them to make themselves manifest, to become actual texts. So, the advertising discourse could be a spot (if broadcast on TV or at the cinema), a press release (if printed in a newspaper), an announcement (if broadcast on radio), and so on. And it is here, therefore, that languages can acquire their own specificity depending on the substances they use (verbal, iconic, gestural, etc.). Every substance of expression will, through the presentation of its communicative potential and its constrictions, tend to give the text certain specific forms.

The following diagram illustrates these several steps:

Narrative structures	Fundamental layer	Semiotic square, axiologies
	Anthropomorphic layer	Narrative programmes, actants, modalities
Enunciation		
Discursive structures	Syntactic component	actors, spaces, times
	Semantic component	themes, figures
Textualization	Relation with forms and substances of expression	

Fig. 1: Visual representation of generative path of meaning.

This model does not aim to present the story of the effective construction of a particular text, but to hypothesize about the theoretical simulacra of this presumed story and to give order to the various semiotic elements of that text. In other words, it is not that a certain author, a musician for example, at the moment of composing a symphony, will have first hypothesized about the profound relationships between pure semantic meanings, then the values, then the actants and modalities, then the actors, the spaces and times, and then themes and figures, and only at the end will they have added in the notes and melodies. This musician will have obviously composed their symphony in its entirety without worrying about generating their text over growing levels of complexity. It is semiotic analysis that, by having to demonstrate the elements in that symphony and the relationships between those elements, projects the model of the generative path of meaning onto the symphony itself, ordering the various elements over various levels of semiotic pertinence.

As we can see, semiotic analysis works backwards, with meaning being hypothesized after its generation. Where meaning starts at the most profound level (the narrative one), moving progressively towards the superficial level of the text, analysis must start from the surface, from the level of the text, and then gradually make its way down towards the more profound levels. At the beginning, analysis finds itself faced with the opacity of the text, its resistance, masked by its manifest existence. Then slowly, by interrogating the text with pertinent questions, this opacity clears and the profound levels begin to appear, revealing the final result of the text's complex underlying workings.

It happens that the more or less implicit 'semiotic awareness' of whoever produces a text does not coincide in any way with that of an actual semiotic analysis. Just as the average speaker does not know the linguistic rules of the language that they use, the producer of a text is almost never aware that a whole series of elements and problems hide beneath the surface of their text that semiotic analysis must, in turn, know how to examine. Consequently, the criteria used, for instance, by the producers of news programmes (the professional rules of journalism and the technical capabilities of television) have nothing to do with the procedures through which meaning is formed and articulated within news programmes (understood as texts). It is after the analysis is concluded that production routines and analysis results can be compared.

II.5 Text and culture

Thanks to this model (whose elements we will illustrate over the coming chapters) it is possible to understand how one text can contain an entire culture. As we slowly move towards the more profound levels of the path, we move towards cultural configurations that are increasingly broad: first is discursive character, then narrative type, before arriving at the great basic semantic oppositions such as life/death, culture/nature, euphoria/dysphoria, and so on. A poem, or even a design product, an urban area or a website appear to be objects whose particular significance is circumscribed within their more or less restricted boundaries. Textual analysis shows us how they are actually specific manifestations of broader configurations. The poem, for instance, will be the particular manifestation of a broader poetic discourse bound to some artistic movement which can, in turn, be understood as a historic reaction to a previous movement within a philosophical, religious, ideological (or similar) battleground. Similarly, a design product such as a sofa is not only an object on which to sit, but the aesthetic manifestation of a social discourse on ways of being: in the lounge, of interacting with each other, of furnishing one's space, and so on. The same thing goes for an urban area, which can be read as a text that holds a complex discourse on the meaning of the city, on metropolitan modernity and so on. Not to mention the website which, by definition, always goes beyond itself in that *vast sea* that is the internet, where discourses are continually being made and interweaving, manifesting general underlying cultural forms.

Recommended bibliography

We can find the origin of the semiotics of text in European structural linguistics (Saussure 1988, 2000; Hjelmslev 1959, 1961, 1975; Jakobson 1963, 1976; Jakobson & Halle 1956; Trubetzkoy 1939), in structural anthropology (Lévi-Strauss 1963, 1966, 1978a, 1978b, 1981, 1983), and in Russian formalism (Erlich 1954; Propp 1958; Shklovsky 1990; Torodov (ed.) 1966; Tomačevskij 1928).
On Saussure: Beividas 2017; Fabbri & Migliore (eds.) 2014; Greimas 1956; Rastier 2015.
On Hjelmslev: Badir 2014; Zinna 2017; Zinna (ed.) 1997; Zinna & Cigana (eds.) 2017.
On signs and inferences see Morris 1938, 1971; Ogden & Richards 1923; Peirce 1931–1958; Sebeok 1976, 1979, 1986, 1991.
First approaches to structural semiotics: Barthes 1967, 1972, 1977, 1982a, 1982b, 1985, 1986 1988; Ducrot & Schaeffer 1995; Eco 1976, 1979, 1994b, 2002; Lotman 1990; Prieto 1966, 1975; Todorov 1982.
On structuralism: Bastide (ed.) 1962; Barthes 1982b; Bertrand et al. (eds.) 2019; Boudon (ed.) 1968; Cassirer 1945; Culler 1981; Ducrot et al. 1968; Greimas 1956; Petitot 1985; Paolucci 2010; Parret 2018; Veron 1996.
On anthropology, folklore, and the semiotics of cultures: Augé 1995b; Barthes 1982a (on Japan); Basso-Fossali 2018; Buttitta 1996 (on Sicilian culture); Courtès 1986 (on fairy tales); Darrault-Harris & Fontanille (eds.) 2008; Descola 2013 (for the ontological turn in anthropology); Douglas 1996; Duranti (ed.) 2001 (a lexicon of linguistic anthropology); Duranti & Goodwin

(eds.) 1992 (on context and culture); Fontanille & Couégnas 2018; Geertz 1973, 1983, 1988 (on interpretative anthropology); Goody 1977, 1987 (on orality and literacy); Greimas 1985 (on Lithuanian myths); Hammad 2017, 2018 (on Arab culture); Hymes 1981; Ingold 2013; Jakobson & Bogatyrëv 1929 (the first structural interpretation of the folklore); Leach 1976; Lorusso 2015 (an introduction to semiotics of culture); Lotman 1990, 2009; Lotman, et al. 1977; Lotman & Uspensky 1984; Lucid (ed.) 1988 (on Tartu school of semiotics of culture); Ochs 1988, 2001 (on Samoan culture); Ong 1973 (on orality and literacy); Propp 1958 (on the morphology of the folktale); Ricci 1994; Sedda 2012, 2019; Serra (ed.) 2012; Sherzer 1983 (on Kuna rituals); Torop, Lotman & Kull (eds.) 2006; Uspenskij et al. 2003; Vibaek-Pasqualino (eds.) 1979 (on ethnic literature); Violi 2017 (o memory and trauma); Viveiros De Castro 2009 (on the idea of multinaturalism); Wagner 1975.

On the dialogue between semiotics and contemporary philosophy: Basso-Fossali 2018; Baudrillard 1988, 1990; Beividas 2017; Bondì 2019; Bordron 2017; Brandt 2017; Sonesson 2017; Cassirer 1923; Coquet 2007; Deleuze & Guattari 1994; Derrida 1988; Foucault 1970, 1972; Goodmann 1968, 1978; Jullien 2004; Klinkenberg 2019; Latour 1993, 2013; Marsciani 2012a, 2012b, 2019; Merleau-Ponty 1966, 2013; Petitot 1985, 2019; Petitot & Fabbri (eds.) 2000; Ricoeur 1974, 1984–1988, 1986; Thom 1988, 1990.

III The logics of narration

III.1 Narration and narrativity

We must not confuse narration and narrativity. Narration refers to those textual products that are understood, in various cultures, as tales (fables, legends, folk tales, etc.), or that tell stories (myths, novels, epic poems, films, comics, plays). Narrativity concerns those characteristics of the tale that are constant, essential, formal, that can be found *in any kind of discourse*, even those seemingly different from tales as we would traditionally imagine them: a philosophical treatise, a work of art, a piece of architecture, a recipe, a ballet, a business document, an advertising image, a city even, and so on. While narration is an intuitive notion that is concrete and changes over time and space, narrativity is a constructed category that tends to be stable within a precise theoretical paradigm: that of textual semiotics. 'Narration' is a term used in everyday language to designate certain works as narrative (and therefore, by definition, certain others as non-narrative). 'Narrativity', however, is a term constructed within the meta-language of semiotics as an explanatory model that brings together a series of different discursive phenomena under certain conditions and from a specific point of view, finding in these phenomena certain formal constants. In other words, narrativity is an *interpretive hypothesis* used to describe the profound structure of any cultural manifestation.

There is a partial overlap between these two phenomena. There is (of course) narrativity in every narration, but there is not always narration where there is narrativity (and this is interesting). It is not necessary for us to point out the presence of narration in particular textual manifestations of, for example, media discourse, and to say that the cowboy in the Marlboro adverts is like a character from a Western. It would be much more useful to locate narrative elements where we wouldn't expect to find them: in branding projects, in strategies and consumer purchasing behaviour, in the restructuring of urban areas, in design projects, in the kitchen, in daily interaction and so on. In this way, it will emerge that the narrative structure of a text guarantees its signifying power and its ability to communicate effectively, contributing decisively to the construction of that fundamental *trust* between the enunciator and the enunciatee, without which textuality has no reason to exist. In short, we tell stories in order to believe.

III.2 Elementary structures of signification

According to the theoretical perspective of textual semiotics, narrativity is a *guided process of transformation affecting one or more subjects* that takes place within any given cultural phenomenon or lived experience. From this point of view, narrativity is the fundamental organisation of every semiotic process of the production and circulation of meaning, i.e. the profound form of human and social experience. The reason is simple: what gives meaning to the world, to our lives and the things that surround us are the transformations, the changes that take place within them. No thing, person or situation acquires significance if it is not in some way compared with what it was before, with what it might become, with some other thing that might take its place. Meaning, before being conceptual, is directionality, projectivity, change. No meaning, no tale comes from remaining static. How can we make sense of what happens? By telling it, placing it in sequence with other events that have happened before and others that happen later. This is why narrativity is the *profound form of our experience*, the grid that places a value on what happens to us, on what we do or go through. It has nothing to do with fiction, with the imagination, with invention as an end in itself. On the contrary, narrativity concerns everyday events and choices, individual and social praxis, the real experience lived by each of us.

As was mentioned in the previous chapter, this transformative process is generally given two different conceptual representations in semiotics, depending on the stratum of meaning that is made pertinent when the text is described: the *fundamental layer* and the *anthropomorphic one*. In technical words, the narrative structures can be described using two different pertinences of the generative path of meaning: (i) the one that is abstract, in which semantic categories are constituted and articulated by internal difference within the *semiotic square*; (ii) the one that is concrete, in which these categories are taken up by human simulacra and the chain of actions come into play, themselves each articulated in *narrative programmes*. In other words, before examining real narrative organisation, we find a draft of the tale in the structures present within the semiotic square. Where there is meaning, this is constructed through relationships of opposition, contradiction, complementarity, and thanks to operations of negation and affirmation. On an anthropomorphic level, this minimal transformation process becomes more concrete, in such a way that the relationships and operations are reconsidered as *procedures of transformation of narrative states*. On this second level, the tale is understood as a succession of states and their transformations, aimed at reaching a final state in which the value at stake is finally acquired.

In the following section, we will illustrate the elementary structures of narrative signification organised within the semiotic square. Following this, we will provide the basics of surface narrative grammar.

III.2.1 Statics: Fundamental relationships

The model of the semiotic square is the *visual representation of the logical articulation of any semantic category* (S). It is a tool used to reveal the category's *internal structure* and the *semi* (s) that this generates. Given that, if we follow the epistemological postulate of structural semiotics – i.e. meaning is constituted by difference (relationships are primary and elements are secondary) – the semiotic square recognises the principle relationships through which semantic categories are constructed in various cultures and eras: the form of signifiers and social values. For this reason, when discussing the square, we talk of *elementary structures of signification*: elementary because they are simpler than others, but also because they are of greater importance. In order that meaning might emerge, in order that even a minimal transformation take place within a given semantic universe (a culture, an interweaving of discourses, a book), it is necessary for at least one of the relationships present in the semiotic square to be made culturally and socially pertinent.

The schema through which these elementary structures of signification are represented is the following, in which the three relationships of *contrariety* (s^1 vs s^2; non-s^2 vs non-s^1), *contradiction* (s^1 vs non-s^1; s^2 vs non-s^2) and *complementarity* (non-$s^1 \to s^2$; non-$s^2 \to s^1$; non-$s^1 \leftarrow s^2$; non-$s^2 \leftarrow s^1$) are activated:

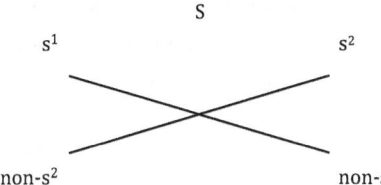

Fig. 2: Visual representation of the semiotic square: the relationships.

This schema illustrates the series of *relationships* that the four terms of the category have with one another. The first is a relationship of *contrariety*, i.e. a *qualitative opposition*, as the two terms at play within the relationship have properties that are opposed to one another. So, for example, in the antithesis /white *vs* black/ the relationship is between elements that in many cultures or social situations are considered opposing colours. Similarly, /tall *vs* short/ is a relationship between opposing elements. The terms of the relationship of *contrariety* are culturally characterised according to their specific intrinsic properties and not because one is the negation of the other, as is the case with the second relationship. We must not therefore confuse *contrariety* with *contradiction*. Here, opposition is *privative* as it places a term with a particular property in a relationship with another term in which that same property is not present. To use the same examples, a privative opposition is /white vs non-white/ (and also /black vs non-black); similarly, /tall vs non-tall/, or /short vs non-short/. If we insert the examples into the schema or (as we would say using technical terms) we lexicalise, we have:

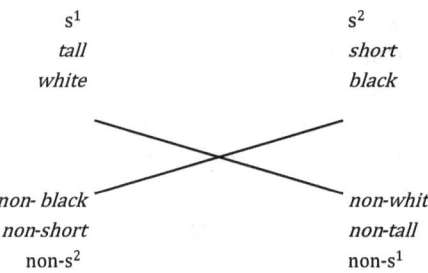

Fig. 3: Visual representation of the semiotic square: an example of lexicalisation.

We can already spontaneously understand that the terms 'black' and 'non-white', 'tall and 'non-short' are not synonymous. Something cannot be white without being black; someone cannot be short whilst not being tall. In a relationship of contrariety, the opposition is between two positive terms that have some specific quality in opposition. In a relationship of contradiction, the opposition is between a positive term (with its own peculiarities) and a negative term that does *not* possess the properties of the term with which it has established a relationship. So, for example, certain social situations dictate a particular colour to be worn ("black is compulsory"), whilst in others the use of a particular colour is prohibited ("do not wear white"), leaving open the possibility of using any other. Similarly, in many traditional cultures loving relationships can be required (i.e. marriage), prohibited (incest), but also non-prescribed, so tolerated (cohabitation) or not-prohibited but opposed (adultery), although of course every society lexicalises these positions using different terms (so the examples given can be modified).

This gives rise to the third relationship represented in the square, that of *complementarity*, which is not an opposition but is still a difference. This arises between those apparently synonymous elements that derive from the intersection between contrariety and contradiction: 'non-white' and 'black', 'non-short and 'tall', etc. In this case, the negative term covers a much broader semantic area than the positive one. In 'non-white' we have all the colours except white, whilst in 'black', there is just one colour. Similarly, 'non-short' can indicate all kinds of different statures except the tallest, whilst with 'tall' there is only one possibility. So, for example, if someone asks me for news on my health, I can respond (using positive terms): "I feel fine" or "I feel ill"; but more often than not I will say (using negative terms): "not bad" or "not great". So, if someone wants to locate themselves politically, they can say they are "right-wing" or "left-wing", but they can also say "I'm certainly not right-wing", "I'm not left-wing", and so on. Euphemisms? A desire to protect oneself? A refusal to respond? Maybe. If we think about it, these are two different thinking dispositions, and consequently, of social behaviour. There are those who reason according to simple oppositions, doling out definitive judgements or giving clear-cut opinions, and therefore only using opposing terms: "if you're not on the right, it means you're on the left", "if you're not short, you must be tall". There are also those who view these clear-cut oppositions as impositions, as extortion, preferring to use less demanding formulas that are, however, better thought-out: "I don't consider myself tall, but this doesn't make me short", "I won't vote for the left but this doesn't mean I love the right". The first attitude constructs simple narrative universes in which everything is black or white, where everyone is either good or bad, in which everyone is either rich or poor. We can see this kind of manicheism in certain popular tales or mass media products. The second attitude articulates more complex and nuanced ways of reasoning and living, in which the gradual and the hypothetical play an important role. The risks inherent in both positions are clear. In the first case, it is oversimplification; in the second, vagueness. The semiotic square, by articulating the two relationships of contrariety and contradiction, and causing complementarity to emerge through their comparison, allows us to avoid these risks, allowing both opposition and difference to exist together in the spectrum of the most elementary virtualities of signification.

Among the most important semantic categories, which we frequently find in diverse discourses and situations, are those that articulate the oppositions *life/ death* (particularly important for the construction of individual semantic universes), and *nature/culture* (fundamental to the construction of collective universes). Human existence is marked before anything else by life and death, but perhaps even more so by 'non-life' and 'non-death', terms that have been valorised differently by different eras and cultures. Think of the delicate questions today regarding euthanasia, abortion, in-vitro fertilisation, the transplant of embryos, and so on, in order to understand the often decisive role played by negative terms in a category: when does 'life' begin? Where is the boundary between 'life' and 'non-life'? And between 'death' and 'non-death'? Similarly, we know that cultures are formed and consolidated when they mark a clear boundary between themselves and their natural 'other'. Every culture places a certain conception of nature outside of itself, thus determining its own formation by detaching itself from that natural alterity ('non-nature'). The majority of those foundation myths of many populations tell of their own human difficulties in marking out the negative terms of 'non-nature' (which is not, or is not yet, real culture) and 'non-culture' (which is, as such, not real nature). The so-called discovery of fire, which also brought us the cooking of food and metalwork, marks this moment of 'non-nature' in many narrations. Thus, in the current context, many brand discourses (from food to tourism, from detergents to cosmetics) propose, in the name of vaguely environmentalist values, a negation of 'culture', a 'non-culture' that gets as close as possible to that Eden-like state of 'nature' that we consider lost forever.

III.2.2 Dynamics: Basic operations

The uses of the semiotic square are not exhausted in this static combination between three types of relationship. In order to fully understand its importance, we must introduce a dynamic dimension. It should be remembered that this model can not only be used for articulating two oppositions and their complementarity, but also for describing the *paths* it is possible to take in order to pass from one term to another. So we must place (paradigmatic) *relationships* next to two (syntagmatic) *operations*:

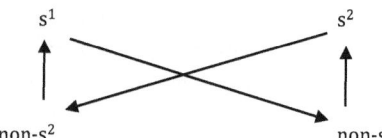

Fig. 4: Visual representation of the semiotic square: the operations.

Here the negations are $s^1 \rightarrow$ non-s^1 e $s^2 \rightarrow$ non-s^2, while the affirmations are non-$s^1 \rightarrow s^2$ e non-$s^2 \rightarrow s^1$. On the one hand, the semiotic square generates terms starting with relationships (and this is their static moment). On the other, it allows us to pass from one term to another (guaranteeing the description of an internal dynamicity within the universes of meaning). This constitutes and articulates the values within a text, discourse or culture, while at the same time outlining some kind of tension towards (and between) these same values and a universe *in fieri*, in constant internal transformation. So, for example, the semantic category that articulates the fundamental opposition between 'nature' and 'culture' is not only susceptible to a static description, but is also given an internal logic concerning the possible transformations between its terms: foundation myths function as a passage from 'nature' to 'culture', through the intermediate term of 'non-nature'. First, they deny 'nature', then they affirm 'culture'; hence the following figure:

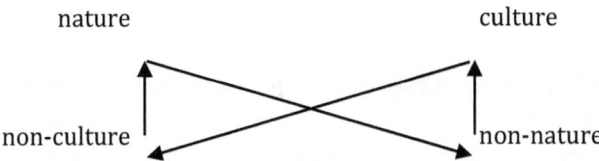

Fig. 5: Visual representation of the semiotic square: an example of the operations.

This schema accounts for the relationships through which each social formation is constituted, generating its own specific natural 'alterity'. It also proposes the paths through which these social formations can be produced, sometimes by negating its state of nature ('nature' → 'non-nature') and affirming that of culture ('non-nature' → 'culture'), or, conversely, proposing returns to the natural through an earlier operation of the negation of the cultural ('culture' → 'non-culture'). So, while many mythical heros (Gilgamesh, Prometheus . . .) follow the first path, many literary heroes (Emile, Robinson Crusoe), along with much of the current media communication, follow the second. By so doing, they direct themselves towards the term they consider positive (as the value to be reached), distancing themselves from the other that they consider negative (a disvalue). The operation of valorisation of the world is, at once, intrinsically logical and potentially narrative.

This is why *the model of the semiotic square is, in short, the description of narrative processes*: if narrativity is the generation of meaning through transformation, this transformation is already present in the passage that goes from a term (such as 'life') and its opposite ('death'), through its precautionary negation ('non-life'). This is also why narrative logic (the primary form of our real experience) collides with Aristotelian logic (formal yet normative). For the latter the principle of non-contradiction (either *a* or *non-a*) is fundamental. For the former, contradiction (reconsidered as negation) is the heart of every story, the motor of every attribution of meaning. The search for the beach on the Mediterranean island that we talked about in the first chapter is in all respects a narrative precisely because it starts from a denial: from the *continuity* of the landscape that is seemingly all the same we move on to the *non-continuity* of cars that suddenly emerge to perception. So newborn babies give meaning to the world around them when they begin to perceive negations. For the child an absence of food or their mother is significant, whilst their presence is taken for granted: it is indifferent, it has no particular value, no meaning.

III.2.3 Complex term and neutral term

Alongside the semes we have discussed so far (*first generation* terms), it is possible to think of other semes (*second generation*) that are constituted when opposing terms find forms of convergence, no matter how fleeting these may be. So, in a semiotic square that articulates the category of sexuality within a mythical cultural setting, the association of 'male' and 'female' generates the term 'hermaphrodite', whilst the association of 'non-female' and 'non-male' generates the term 'angel'. The Christian religion responds to the disturbing figure with dual sexuality, typical of pagan cultures, by inventing angels – doubles without sexuality, ambiguous because they are ethereal. Hence the following schema:

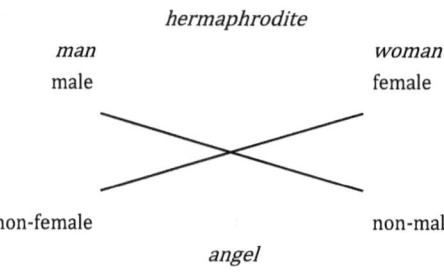

Fig. 6: Visual representation of the semiotic square: complex and neutral terms, an example.

The term that unites the opposing semes is a *complex* term, while the term that unites the sub-opposing semes is a *neutral term*. The complex term constitutes an enrichment of meaning, if not a return to a kind of original indistinction (a bit like the way the hermaphrodite, according to Plato, preceded the distinction between the sexes). The neutral term often tends towards non-sense, semantic indifference: denying both terms of an opposition ('non-tall' + 'non-short') often leads to the omission of the entire semantic category (stature is no longer pertinent in that given context).

Old rhetoric referred to complex terms as *oxymorons*, describing expressions such as the 'young-old' or the 'stupid-intelligent' and so on. But everyday language is also full of these kinds of terms: 'tepid' is the neutralisation of the opposition between 'non-hot' and 'non-cold', indicating something that is neither hot nor cold. Myths are a boundless reserve of complex terms (demi-gods, animals/humans, etc.), but advertising also makes extensive use of them, using them to reconcile properties that are normally perceived as opposing (cheap *vs* elegant, robust *vs* refined, artificial *vs* natural). These choices are sometimes valid, though more often than not they are risky: the mythical demi-gods often meet a sticky end, and with them the sphinxes, minotaurs, sirens, and so on. Behind every complex term hides a neutral one, with something that is sold as both cheap and elegant often ending up being neither one nor the other.

III.2.4 The constitution of axiologies

A final observation refers to the issue surrounding the constitution of values. The semiotic square has another role: that of producing *axiologies*, value systems, valid in specific socio-cultural contexts. How can it be, for example, that in certain cultures 'black' clothing is considered positive, whilst in others it is viewed as negative with its opposite, 'white', as positive? How is it that in a lot of ancient societies, 'war' is a positive value and 'peace' a negative one, whilst in modern cultures the opposite is true? The formal mechanism to explain all this is rather simple.

In order for the articulation within semantic categories (visualised using the square) to produce value systems, the terms must acquire a significance that is at times positive and at times negative. Thus, for example, in the square that takes into account the chromatic opposition between 'white' and 'black', each of the four terms can acquire a positive or negative value depending on the cultures, the universes of discourse, the forms of life, the texts in which the category is used. This is the case with bridalwear, where, in Western culture, the opposition between the two contradictory categories of 'white' and 'non-white' is pertinent, with the former being valorised positively and the latter, negatively. When it comes to signalling mourning, we know that the pertinent term in Western cultures is 'black' (valorised nevertheless negatively as it indicates death), whilst in certain Eastern cultures the pertinent term is 'white' (which acquires a negative value for the same reason). We can say, then, that in order to generate values (in different ways depending on cultural context), it is necessary to introduce another semantic category to the semiotic square representing the logical articulation of the same. This new category is known as *timic*, and it distributes the opposition *euphoria* vs *dysphoria* between the various terms. The homologation between the seme 'euphoria' and a particular term (such as 'white'), produces the value 'white = positive', whilst the inverse operation, so a homologation between the seme 'dysphoria' and another term (such as 'non-white'), will produce the value 'non-white=negative'. In this case we will have:

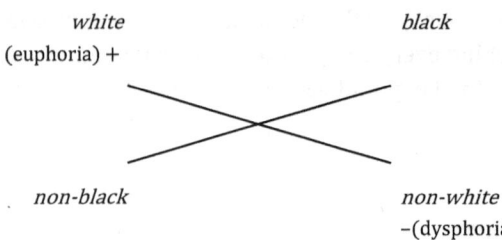

Fig. 7: Visual representation of the semiotic square: the constitution of axiologies, an example.

It happens that, within semantic universes, discourses, social contexts, and so forth, certain social values ('justice', 'wellbeing', etc.) or certain individual values ('eros', 'self-affirmation', etc.) are generated from the way in which semantic categories are related to one another, from the way in which these values enter into a relationship with opposing, contradictory or complementary terms within the specific semiotic squares that are pertinent in those contexts.

But what does the timic category consist of exactly? It is the spontaneous semanticism linked to the way in which human beings perceive themselves, their own bodies, their surrounding environment. From these things come pleasure and displeasure, pain, annoyance, repulsion, attraction. Before all of that becomes the object of cognitive reflection (which transforms attraction into interest, or repulsion into disinterest), even before understanding whether and how the world is important, useful, or interesting for us, we must first understand the fact that things, people, places, situations can cause us annoyance or pleasure, because they are dysphoric or euphoric. Thus, for example, it is widely known that the cry of a newborn baby should be interpreted as a sign of general dysphoria, and not as the signal of a specific physical or psychological pain. From this point of view, it seems obvious how the difference between useful objects and those that we simply like, between functional spaces and environments that make us feel good, and more generally between instrumentalism and aesthetics, cognition and pleasure, is fundamental for the construction of many cultural configurations.

Furthermore, the timic category is articulated within the semiotic square. By expanding the opposition between the opposites of 'euphoria' and 'dysphoria', we find 'non-euphoria' and 'non-dysphoria', along with the complex ('diaphoria') and neutral ('adiaphoria') terms. These last two are particularly useful because they allow us to explain the existence of minimal urges towards something or someone; urges that are not yet distinguished in terms of positive or negative, attractions or repulsions. Often, for example, the media work to obtain precisely this kind of effect. They aim to attract our attention, to create tension towards an object or situation that is without that attention, that tension, to create a distinction between positive or negative.

III.3 Elements of narrative grammar

It would be reductive to attribute narrativity exclusively to the operations of the semiotic square, negation and affirmation. Very often, it happens that tales (such as myths) describe certain content (such as a situation of war) and end with its opposite (peace). In these cases, we say that the tale is the passage from an *inverted content* (which lies at the beginning) to an *established content* (found at the end). In this way, the tale's 'message' lies neither at the beginning nor the end, but in the passage from one to the other, i.e. in the *process of transformation*. As well as considering this double logical transformation, negation and affirmation, it is necessary to understand if and how these operations are undertaken by anthropomorphic subjects. In terms of the sphere of individual and collective experience, these operations are no longer logical movements but *precise moments in a story*. At the anthropomorphic level, the tale is thus configured as a progressive transformation of states in which subjects, objects and most importantly, values, are at stake.

Let us first try to understand what a 'state' is. It is a relationship of conjunction and disjunction between two *narrative actants*: a Subject and an Object. The transformations are passages from one conjunction to a disjunction, or from a disjunction to a conjunction. They give rise, therefore, to a disjunctive transformation and a conjunctive transformation. In the syntax of any given phrase, the subject and the object are always present, though sometimes implicitly, in the action expressed by the verb. In the same way, within a tale these two basic actants – Subject and Object – are always present. This is why we can say that a tale is the same as a phrase: there is a process (doing), a number of protagonists (being), further additional elements (other processes, other actants). The tale is not, as is often maintained, a simple series of actions, but more precisely, a *non-casual succession of transformations of states* – at times conjunctive, at times disjunctive – *that aims for a final result*.

III.3.1 Subject and object

It should be made clear that the Subject and the Object are not, respectively, individuals or given things that entertain a relationship. They are *terms*, and they define each other only in their reciprocal relationship. There cannot be one without the other: the Subject is that narrative element that is conjoined or disjoined with the Object (so not necessarily a human character). The Object is that other narrative element that exists in its disjunction or conjunction with the Subject (and is not therefore a thing in the literal sense of the term). Both are *actants*, syntactic elements through which the semantic forces present in a particular tale are articulated. This means that *no subjectivity* – individual or collective, social or institution – *exists without some kind of intentionality*, a *directing oneself towards* something external to it, be it a thing, a person or an idea. In the same way, in the universe of meaning (which is not the physical universe but the social and cultural environment: the *semiosphere*) there is no external objectivity, independent, extraneous to us and obstinately mute: *objectivity exists always for a subject*.

III.3.2 Subject of doing and subject of being

In a tale there are two types of Subject: a Subject of Doing, which sets in action the transformations, and a Subject of Being, that is conjoined with or disjoined from the Object. In a tale, two figures can coincide: it might be that a princess (Object) is kidnapped from a king (Subject) and he himself goes to rescue her. More often, it happens that a king (Subject of Being) sees his daughter (Object) kidnapped, but that another hero (Subject of Doing) leaves to get her back. This means that we have to distinguish between the two forms of subjectivity.

On the basis of this distinction between two types of Subject, two diverse dimensions of meaning take shape. For the Subject of Doing the *pragmatic dimension* is established. Meaning is produced by the events that happen, in the actions carried out by the subjects, in their behaviour, in their human and social praxis. For the Subject of Being, the *passionate dimension* is established. Meaning is constituted thanks to the subject's affects, which agitate the states of conjunction and disjunction with the object (timic moments that are, respectively, positive and negative, of pleasure and displeasure, euphoria and dysphoria). The states, seen from this point of view, are in no way static: passionate processes are carried out within them that are constitutive phenomena of narrativity as much as actions.

III.3.3 Objects and values

The Object – material or human – is not important in itself, owing to its properties or intrinsic features, but is important for the *value* that is inscribed in it and attributed by the Subject. A Subject, in fact, can go in search of an Object with which it wants conjugate for very different reasons. Thus, the hero can search for the princess in order to obey the king (value: 'sovereign authority'), to climb the social ladder (value: 'money'), because he is in love (value: 'eros'). *The Object is always an Object of value*: a value that is more or less concrete, more or less abstract, according to the situation, but always linked to a subjective experience. Identifying the Subjects and Objects of a narrative structure will, for semiotic analysis, dignify the values at play in the experience. Thus, every object emerges as significant and becomes desirable thanks to the attribution of value by a Subject.

All of this has consequences, for instance, when observing the consumer universe. Despite every form of philosophical ontology, no product can expose itself to the market as it is, with only its own materiality, its own instrumentalism. Products are always charged with meaning and value, at times by the producer, at times by the consumer, and at others by both, or by the society or culture in which they circulate. There is no *before* for the product as it is (take, Coca-Cola for example), bestowed with a number of factual properties (taste, colour, temperature, etc.) and certain specific functions (to quench thirst), and *after*, the eventual attribution of meaning, the value surplus that the brand uses to its advantage (feeling symbolically similar to young Americans, adhering to a social and spontaneous lifestyle, and so on). Quite differently, *the constitution of the product*, as a human and social object, *coincides with its subjective valorisation*: quenching one's thirst with Coca-Cola means leading a certain kind of lifestyle. *The world of things and the world of meanings are integrated with one another.* In this way, the product's function, the practicality of the object, even its material features, are *effects of meaning* that give rise to an operation of valorisation: sometimes explicit, sometimes hidden, other times forgotten, but always present. If we invert our point of view, no signified or value exists as an entity in itself, a sort of 'symbolic capital' managed by the brand with its communicative alchemy. In order to exist, meaning always needs its own empirical manifestation, its own historic and social tangibility. In order to exist, value needs its own inscription into any object of the world (the bubbles in the Coca-Cola are both a chemical principle and a socio-cultural value). This is why, in semiotics, we prefer the term *valorizations* to values. Valorizations are the actions through which Subjects attribute meaning to an Object that in one fell swoop produces the meaning and the Object, the value and the product, the signified and the signifier.

III.3.4 Actants and actors

We mustn't confuse *actants* (which appear on the anthropomorphic level of narrative structures) with a tale's actual characters, or *actors* (which appear on the level of discursive structures). In the simplest cases, every actor corresponds to an actant, as in the tale in which the hero is the Subject of Doing who wants to conjoin the prince-Subject of Being with the maiden-Object. But it can also be that just one actor personifies two different actants – like when the Hero is at the same time the Subject of Doing and the Subject of Being, or when someone, going in search of themselves, occupies all actantial positions. The last possibility is that a single actant is personified by several actors. Think of the tales in which three brothers (Subject of Doing) set out together in search of their vanished parents (Object of value). In this third case, in which there are many actors whilst there is only one actant, we refer to a *collective actant*: a social class, a football team, a family, an army, a company, act like any other subject, they are searched for as objects, etc.

III.4 Programmes, modalities, identities

A series of actions is not enough for narrative structure, which requires some actants to set a sequence of states and their transformations. Such a sequence is not casual. It is *guided*, tending toward a final state, that is the conjunction. The Subject, in order to realise itself, must be able to make an Object of value their own. Or, if it is a Subject of Doing, it must ensure that the Object become conjoined to the Subject of Being. As such, the narrative sequence has a precise target, a specific aim, a result it must attain and thanks to which, at the end of the story, a particular Subject is realised, becoming a real Subject, someone who is conjoined (or caused someone else to conjoin) with their own Objects, or rather, their own values. In other words: the target is someone who has acquired *an identity* ("I am someone who has managed to attain what I wanted"). The subjective identity is the result of a successful narrative sequence, the consequence of the completion of a *narrative programme*. Thus, identifying a tale's structure is equivalent to reconstructing the narrative programmes that allow for its development, starting with specific stakes (the search for certain values) and moving toward a particular objective (the conjunction with those values). A narrative programme (or, NP) is defined as *the collection of operations that a Subject of Doing sets in motion in order to ensure that the Subject of Being can be conjoined (or disjoined) with its Object of value*: the series of stages through which these Subjects must pass in order to reach the final result.

III.4.1 Modalities and modelling

Firstly, in order to carry out the programme of the search for the Object of value, the Subject of Doing must be enabled to do it. In order to *carry out a particular action it is vital that they are capable of doing it*. It is first necessary for them *to have to* or *to want to* do it. As with linguistics, the narrative process is modelled differently depending on the *modality* with which the Subject is made capable of carrying out the action. A Subject according to *will* is one thing, a subject according to *duty* is quite another. In a tale, for example, we must distinguish between these two cases: someone who wants to carry out a particular task, who finds within themselves the reasons for their own actions, who fights in the name of their own ideals; and someone who is forced to do so, who obeys orders, who adheres to superior laws, etc. And again: one thing is a subject who is able to do something, or who has permission to do it, but a subject who is able to do something, who knows how to do it, is quite another.

There are four narrative modalities: *having-to* and *wanting* (referred to as *virtualising*), being-able and *knowing-how* (*actualising*). Strictly speaking, these are verbs that directly predicate not a descriptive content but other verbs (in traditional grammar they are known as 'modal' verbs), radically transforming meaning. They are verbs expressing action (*doing*) or a state (*being*) added to these: have-to, want, be-able and know-how to do; have-to, want, be-able and know-how to be; and their respective negations – not-having-to do, not-wanting to do, non-be-able to do and so on, either for doing or for being. Furthermore, each modality can constitute the predications of another, and there are almost infinite combinations of this: having-to-want, having-to-know-how, having-to-be-able, wanting-to-be able, wanting-to-know-how, etc. In this way, the analytical possibilities of narrative processes become vast and we will see their utility, for example, when it comes to describing pathemic phenomena or communicative pacts.

In every canonical tale, in order to pass to the act, in order to operate the transformation, the Subject of Doing needs to first acquire a having-to or a wanting (becoming a simple *virtual* subject), and then a being-able or a know-how (becoming an *actual* subject). Depending on the acquired modalities (or those not acquired), the transformation will take place – or not, the conjunction of the Subject of Being with the Object of value will be carried out (or not), and the Subject of Doing will, in turn, become a *realised* subject. Thus, a typology of subjects (virtual, actual, realised) is created, or rather, a series of *progressive assumptions of identity*, of degrees of semiotic skill, and therefore, existence. The Subject, as we have seen, is constituted within the narrative process thanks to its valorising relationship with the Object. We will now see that this identity is never given once and for all but is progressively outlined, transforming through the modalities the Subject possesses (or is given).

Narrative subjectivity is adjustable. It is not constituted by an accumulation of properties, by specific entities present in a certain character ("I am like this-and-like this"). On the contrary, subjectivity varies according to context and is enriched or impoverished by the wants and the knows-how, by the have-tos and the be-ables that it acquires or loses, that it manages to obtain or not. It is a question of *forms of life*. There are subjects who have the wanting but not the be-able, and who therefore are unable to ever fully realise themselves. There are subjects who, as soon as they have a certain being-able, start looking for another and never actually get to the real action, so that being-able in itself becomes (for them) an Object of value. There are subjects who have the know-how to do, but not the being-able, and therefore they are never in a position to act. Once more: there are subjects who have the being-able but not the wanting, who ask themselves what is really worth doing. Subjects who put wanting and having-to together, in keeping with certain Protestant ethics, also exist. There are Subjects who do without being-able, and improvise, causing damage to others. There are also Subjects who progressively lose interest in the Object as it slowly becomes more accessible, and who question what they are doing. In addition, there are Subjects who are not able to do because someone forbids it, causing them to suffer. And there are Subjects who are able-not-to-do, and who use their lack of action as a weapon.

This is why, in narration, subjective identity is at once pragmatic and pathemic: the calculation of subjective skills that are local and temporary, casual or searched out, causes, through a series of programmes of action, an emotional outburst; or (which is the same thing) the quick succession of affective build-ups sets operative programmes in motion.

III.4.2 Basic programmes and instrumental programmes

We see, then, that the narrative programme (NP) which aims for the conjunction (or the disjunction) of the Subject of Being with the value inscribed in the Object, is a *basic* NP and it is accompanied by one of more *instrumental* NPs, which are used by the Subject of Doing to find the modalities through which they can pass to action. The narrative modalities are therefore configured like any other *values of use*, which are also inscribed in Objects, and can also be reconfigured in various ways that are more or less concrete, more or less ideal. The monkey that wants to eat the banana (basic NP) must first acquire the being-able-to-do (a stick, for example) that allows it to knock the banana down from the tree (instrumental NP). While the banana is inscribed with the basic value ('nutrition'), the stick is inscribed with the instrumental value ('being-able-to-do'). In the same way, if someone wishes to seduce their beloved object (basic NP), they must make themselves desirable, agreeable (instrumental NP). They will therefore equip themselves with the means (clothing, cosmetics, hygiene, etc.) that allow them to become this way, before setting the final programme of conquest in motion. A large number of social discourses are organised according to this logic, which is, as we know, the logic of common sense, the profound articulation that regulates our lived experiences.

It is easy to understand how the junctures between narrative programmes can reach very sophisticated forms of complexity. For example, the suspension of the basic NP leads to an activation of a series of instrumental NPs, which, once completed, allow us to return to the initial basic NP and bring it to a close. It often happens, however, that attempts are made to set basic NPs in motion without passing through the instrumental NPs (action is taken without the right skills), or that the NPs are abandoned in favour of purely instrumental NPs (becoming equipped with a useless skill). So, while the acquisition of a skill must *in principle* always precede the passage to action, there is no rule that this will *actually* be the case, that the subjects will desire the events to unfold in that particular order. Very often, we want useless things, we equip ourselves with skills for which we have no use, and then, once we are conjoined with these, we set to doing things that have never interested us, for which we felt no need.

Here we find a fundamental phenomenon regarding the relationship between the economic sphere and that occupied by the media, between the market and mass communication, and that involves brands. It is the widely discussed problem of the overlapping of desires and needs, of the induced (or otherwise) arrival of the former to the detriment of the latter, of the alteration of consciences that transforms citizens into consumers, individuals into clients. If someone finds themselves acting on the basis of skills that they did not want, where did that desire come from? How and to what point is it possible to impart it to them?

III.4.3 The Addresser

How does the operational Subject acquire the first modality? How is it that a certain character in the tale becomes the Subject charged with carrying the main narrative action? We must presuppose the presence of a *third actant* who confers the first modality on the Subject of Doing necessary for the action to take place (be it wanting or have-to), placing them in the position of a virtual Subject who aims (for themselves or the Subject of Being) for some Object of value. This third actant, the *Addresser*, is the one who, hailing from another dimension (and therefore transcendent by definition, in terms of the given narrative universe), transmits the values that they carry to the Subject of Doing. For example, in an advertising spot, a father tells his son to practice a particular sport, making him rise to the condition of the story's virtual hero. He can do this in the name of his own personal authority, providing the boy with a having-to-do, or in the name of common interests, imposing a wanting-to-do. These are two very different modalities. What makes the wanting-to possible when it is strength or social hierarchy (be it religious, economic, family, etc) that constitutes the having-to? In the name of which principle do desires make their way into others? The response is a complex and extremely delicate one that can vary considerably, depending on context or situation. Yet it allows us to glimpse how *the Subject is never a monad*, a single individual, but is configured as a social actor who, before beginning their narrative journey, experiences controversial or contractual relationships with other social subjects within hierarchical or equal inter-subjective relationships that provide the subject with having-tos and wantings, obedience and alliances. We all have our own Addresser, be they a divine entity, an institution, a community, a particular individual, a mythical or imaginary being, a family, a star, a political party, a brand, and so on.

While the tale is a closed structure (with a beginning, development and end), the figure of the Addresser keeps it open and in contact with a different semantic universe from which values come and to where the Addresser will return at the end. While the Object is a something in which subjective values are inscribed by a Subject, such values come to the Subject from the outside. This means that every tale is in constant, necessary contact with other tales, with other stories, where different characters (for example, a mother and child) embody the same roles (for example, that of the Subject) or vice versa, the same character (for example, a doctor) has different actantial roles (first Subject, then Donor). The relationships between the given narrative universe and the transcendent universe are managed entirely by the Addresser: a figure who mediates between different environments and contexts, and who, at the outset, confers the values at play upon the Subject, and at the end of the same story, judges the Subject's work according to the values brought into play at the outset. The Addresser is therefore both an *instigator* and a *judge*. As actants they can be represented by the same or a different actor.

It is fairly evident how this figure of the Addresser is even more important than that of the Subject, since the values that the Subject vows to reach through their own programme of action depend on the Addresser. It is the Addresser, rather than the values in themselves, who determines the values of the Subject, their social weight, their cultural hierarchy, the importance that should be attributed to them: in short, their *worth*. Having a good Addresser means already being in a good position, while not having one is a significant disadvantage. Politicians know this well, alternately proposing as their Addressers the People, the Country, the Church, the European Community, even NATO, indirectly transferring the authority that comes from any of those realities onto themselves. Brands are also well aware of this, equally concerned with equipping themselves with

Addressers who are glorious or otherwise, legendary or otherwise, acceptable or otherwise, and they are equally ready to put themselves forward as Addressers for their potential consumers. Brands receive value from the outside, and transfer it to others. This is a significant part of their work, and it places them, for this very reason, in a narrative position that aims to imitate the sacred sphere.

III.5 Canonical narrative schema

What has been said regarding narrative structures can be summed up in the *canonical narrative schema*, a model with four stages that can be used to analyse every aspect of narrativity, be it the tale itself, considered such by a particular culture, or an abstract structure of other genres of discourse (from philosophy to cookery, from consumption to daily life, etc.).

The central point in any narrative structure is the *performance*, the act that, if successful, leads to narrative transformation. An act that allows for the passage from an initial (often negative) state to a second (often positive) state. In fairy tales this moment corresponds to what folklorists refer-to as the Struggle Function, known by mythologists as the Decisive Task, as it leads to conflict with an Antagonist, i.e. an Anti-subject who carries out an opposing narrative programme within the same tale. The performance is generally understood as the action (or series of actions) thanks to which the Subject carries out their basic narrative programme. But the success of this act is not a given. The performance is therefore a *test of the Subject's abilities*, of their being-able and having the know-how to do, against the abilities with which the Anti-subject is, in turn, equipped. But, fundamentally, more profoundly, it is a kind of test of the social purchase of its values, of their real cultural consistency, of their *worth* when faced with the opposing values brought by the Anti-subject (which are also indirectly put to the test).

Though the act of performance is necessarily present in every narrative structure, it is not the most important. It is essentially the result of what goes before it and the cause of what follows. The Subject that carries out this fundamental transformative act, as we have said, must first be placed in a position to do so: they must acquire a *competence*. It cannot be possible to carry out a certain action without knowing-how or being-able to do so. The action of the performance (which corresponds to the success of the basic NP) must therefore be preceded by other actions (articulated in the instrumental NP) that are consistent with the acquisition of the skills necessary to carry out the performance of Struggle with the Anti-subject (the being-able-to-do and know-to-do). In fairy tales, this moment is often presented as an action of Receipt, often in the form of a Gift, of that magical agent that allows the hero to resolve the initial Trickery.

In myths, this is the famous Qualifying Task through which the hero will gain possession of the objects, information or allies that will allow them to acquire extraordinary, supernatural, mythical powers. More generally, the *acquisition of competence will consist in that action or series of actions thanks to which the Subject is placed* (or should be placed) *in a position to pass to action*. Such acquisition is, by definition, almost never peaceful. If, on occasion, the 'magical agent' is the result of a Gift, or the presence of a Donor is foreseen, very often there is a trial to overcome, a task to carry out against an Opponent – another narrative actant that blocks the completion of the instrumental NP, the possession of being-able and know-how. In advertising terms, it is not just the seduction technique of the beloved object (basic NP) that is complex, but also the acquisition of the shampoo with which to make oneself beautiful (instrumental NP).

Yet this is not the entirety of the story's structure. In every tale there are values at play. This is why the two *pragmatic* moments of the narrative schema, the ones in which the actions are carried out, are framed by two *cognitive* moments, where the issue of narrative values is at play. The first of these is the *manipulation*, an initial element in every tale in which the Addresser and the Subject stipulate a contract. On the basis of this contract the Subject acquires a wanting or a having-to. Taking the example from advertising, the moment of manipulation can be found in the moment in which the brand insinuates to the consumer the reasons for their own act of acquisition. As we have hinted, the provision of having-to is easier as it derives from the use of force, or in any case, from the demands of some authority ("you must do this"), but it is also more ephemeral because one can rebel against authority ("I don't want to do it"). The provision of wanting is much more delicate as it requires a procedure of persuasion of the Subject's interior conscience, a real movement to convince them of the validity of a particular value system ("you want to do this"). This is why the moment of manipulation is also that of the contract, of that implicit *fiduciary* agreement between Subject and Addresser. In order to adhere to the Addresser's values, to want those values and make them their own, the Subject must first believe the Addresser, have trust in what they say and promise ("if you do this and this, in the name of these values, you will receive a reward"). Once this trust is obtained and the contract stipulated, the manipulation will be much more efficient than that obtained through the transferral of a having-to. There is no longer an authority that imposes itself. Rather, there is an *assumed* wanting, a system of values that is made the Subject's own, in the name of which they proudly and euphorically act (and suffer) – to the point of making the Addresser disappear.

The second cognitive element of the canonical narrative schema is the *sanction*: the final moment in the tale in which the Subject, the performance now over, returns before the Addresser (who has reappeared in the form of the same or another actor) and submits their actions to the Addresser's judgment. If the sanction is positive (corresponding to the values agreed upon in the initial contract) the hero will be transformed. If the sanction is negative, s/he will fall once more into the anonymity typical of non-subjects. In fairy tales, this is known as the Wedding function, the moment in which the hero, who has returned home clashes with the false hero who wanted to take his place. In this way, the hero can show his people that he defeated the antagonist and be properly recognised as a hero, also receiving his reward (marrying the princess, half of the kingdom, etc). He will be transformed into a very different person to who he was at the beginning (if he was a farmhand at the start, he will be a king at the end). In myths, this moment is known as the Glorious Task, as the mythical hero, now his quest is over, no longer needs the superhuman agents he required to carry out his mission. Nor does he need to undertake any restorative actions against his sworn enemy, although he will need to put himself to the test once more in order to obtain the social recognition of his mythical essence: the institutionalisation of his transformation.

The moment of sanction is generally one of reckoning with what has taken place. An attempt is made to understand whether the assumption of values by the Subject has been conducted. It is a judgement of the Subject's performance ("you did well"), but more universally of the skill that he has acquired ("you had the ability to do it"), and, indirectly, of his value system ("you behaved correctly"). What is now at stake is the contract stipulated between the Addresser and the Subject, according to which the Subject would have the right to a reward for a successful mission. Once more, it is the figure of the Addresser (first instigator and now judge) who emerges in all their importance; not so much in the name of the singular events that have taken place over the course of the story, but in terms of the cultural substance that supports them, the social value that they have or could have, in terms of what they signify. It is the Addresser who provides a destiny, who reassures us of the profound meaning not only of our own entire existence, but of all those actions of ours that can create a system within a well-executed story.

The canonical narrative schema can therefore be visualised in the following way:

manipulation				sanction	*Cognitive dimension*
	competence	performance			*Pragmatic dimension*

Fig. 8: The Canonical Narrative Schema.

and requires certain fundamental clarifications.

III.5.1 Narration and reasoning

The canonical narrative schema is placed on a level of pertinence concerning meaning in a very profound sense. This means that not all moments within the schema must necessarily be present in a given text – be it literary, advertising, gastronomic, spatial – nor present in a given lived experience. In popular fairy tales, it is common for each of the narrative moments to be recounted one after the other. While in other kinds of text, or other narrative situations, it is likely that some states of action can simply be alluded to, such as with backstory or other forms of narrative ellipsis. In many stories, the moment of manipulation, as fundamental as it is, is never told, though it is nevertheless understood through the hero's actions, through the reasoning that leads them to follow particular programmes, to identify certain characters as enemies and others as friends. In other stories, however, it is the performance that is not told. For example, if, in a film, we see that a certain character is given a sword that makes him invincible, it is not necessary for him to be shown in the next scene using it against the Villain: we already know he will win, and we can move directly to the sanction. Thus, *narration is a form of masked reasoning*. Apparently, I only tell one story, paying little attention to certain aspects; but I have, nevertheless, included these neglected aspects in my narration: I have hidden them, not actually omitted them. If I tell the story of a man who did not carry out his task properly (who did not execute his performance well), I am indirectly alluding to the fact that he does not have the right skills to do so. I am therefore calling into question the moment of his skill acquisition or that of the manipulation, or perhaps I am directly attacking his Addresser. Similarly, if I depict someone reaching a state of ecstasy as they eat a certain brand of ice cream, apparently demonstrating the positive moment of the product's consumption, I am effectively alluding to the productive capacities of that brand, to its skill in providing pleasure.

III.5.2 The narrative presupposition

How is the mental operation that allows us to reconstruct an entire tale from a single fragment possible? Through what kind of reasoning do we move from the single variant to the unique invariant model? This model must belong to the interpreter's culture of reference. Once this has been established, it must be made clear that the possibility of starting from one moment in the schema in order to rebuild all others is *presupposed*. This means that it is implied that you could move backwards through the schema but not forwards. The completion of a performance assumes the acquisition of the competence, the carrying out of a sanction assumes a performance, even if neither of these are recounted. But the opposite is not true: acquiring a skill does not necessarily imply a passage to action. A Subject can know how to do a certain thing but this doesn't mean they can actually do it. A performance can take place but does not necessarily imply a subsequent sanction. A Subject can have carried out their task but not be judged. Many existential dramas play out in this way: there is a will but not the consequent possibility; the skills have been acquired but there is never an opportunity to use them; a good job is done but no one recognises it. *Knowing how a story is going to end is the only way to reconstruct the meaning of the recounted events.*

Moving backwards in narration thanks to presupposition is an irrefutable mechanism within a certain culture: if I have acquired a car, I will first have to find a way of getting the money, otherwise it will not be an acquisition but another form of appropriation of that product (gift, inheritance, robbery). In this sense, the presupposition is a very powerful and effective narrative/argumentative form. However, its necessity risks banality ("it is obvious that if you do this thing well, it is because you possess the skills to do so"). Conversely, the story's onward movement is entrusted entirely to imagination, to a fantasy masked by implication ("it might happen at this point that . . . "). Though weak in terms of argumentative rigour, this mechanism of implication is rich in terms of creativity and imaginative allusion. Advertising is well aware of this, particularly when it uses irony to construct a complicity with its receiver/consumer: showing someone using a certain perfume can make you think, for example, of its seductive capacity, and so cause you to imagine marvellous romantic adventures – unless it's the wrong perfume . . .

III.6 Polemics and strategies

The narrative schema, however, must be doubled. Within every tale there is a *polemical structure* in which at least two basic narrative programmes meet: that of the story's Subject and that, formally homologous, of their Anti-subject. If the first of the two Subjects is considered a 'hero', it is because whoever tells the story tends to share their values. In the same way, the Antagonist (the 'baddie') is only such because their values are rejected by the narrator, and considered anti-values. But beyond the *value perspectives* through which the stories are ideologically interpreted, it is beyond doubt that within each tale there is a contrast between the two programmes. As such it is necessary to hypothesize the presence (even if implicit) and transformations of two Subjects, two Addressers, two diametrically-opposed value systems (inscribed in the same Object or in two different Objects). Thus, *meaning is found in conflict*, and the semantic oppositions occur as contrastive actions. Every programme is also, therefore, an anti-programme. Every act has meaning because it is opposed to another; every value is such because it differentiates itself from an opposite value. Absolute meanings and values do not exist, only relational meanings and values. To praise the autonomy of a value is to forget, or suppress, the conflict that has led to its emergence in respect of the opposite value.

III.6.1 Conflicts and identity

There are at least two consequences of this. The first is that *to tell a story means always and necessarily taking a position* in favour of one of two Subjects. This has a semiotic reason: assuming a narrative perspective means bringing into play values and anti-values, Subjects and anti-Subjects, signifieds and opposing signifieds. It is impossible to assume a neutral point of view, detached from the values of the two Subjects. Enunciating a story means claiming the values assumed at the moment of its narration. The second consequence regards the issue of subjective identity, individual and collective, personal and institutional. If the Subject is continually constructed and transformed within a story, it is not only because of their goals in terms of the Object of value, which cause them to be modelled in different ways – be it virtual (when they acquire having-to or wanting), actual (when they require being-able and know-how) or realised (when they finally reach the Object). There is another reason that involves intersubjectivity: each Subject is such because of their Anti-subject, being formed and realised through their relationship with their Other. They are shaped and strengthened through their struggle against the Enemy, their own ideals are reinvigorated or weakened depending on the weakness or strength of the Antagonist. *Without conflict there is no real transformation, and therefore no real identity.*

The Subject's identity is, therefore, two-fold. On the one hand, it is constituted in the narrative path (identification of the value, inscription of the value in the Object, desire for the Object, search, conjunction, etc.) while, on the other, it is affirmed through the dialectical conflict with the Anti-subject, with everything that this conflict carries with it in terms of strategic thought (as we will see).

III.6.2 Alterity and simulacra

This is where the fundamental role of strategies comes in. If the Subject and the Anti-subject meet at the moment of the performance, their respective action strategies presuppose the presence of the other. Thus, the Subject, before encountering the Anti-subject, must construct a *simulacrum* of the Anti-subject for themselves; they must imagine their opponent's possible moves, adapting their own to those they assume the Anti-subject will make. The Anti-subject must set in motion the same strategy, constructing a simulacrum of the Subject in an attempt to predict their moves. The global *strategies* of the basic NP are accompanied by the localised *tactics* of the *substitute* NP, paradigmatic programmes set up to remedy or anticipate the actions of the other (whether real or presumed).

III.6.3 Multiplication of subjectivities

Strategies and tactics multiply the narrative actants, increasing the importance of the cognitive dimension of meaning. In contrast to more traditional theories which consider the tale a series of actions ordered according to causal and temporal criteria, the idea of narrativity leaves space for cognition and passion. *No action is possible without knowledge and affectivity*. The strategist is the person who raises the issue of conflict between their own narrative programme and that of their enemy or, rather, the one they assume to be their enemy. Hence the historical importance of spies, actors charged with spreading tactical knowledge about the enemy's skills; and, historically earlier still, the importance of coming into contact at an opportune moment with the information useful for constructing one's own war capacity. Furthermore, the strategist knows that the enemy reasons in the same way and must therefore offer a particular image of themselves, a particular simulacrum of an action plan that compromises their enemy's imagined programme. Hence the presence of counter-spies, and more often, the dirty game involving many spies. These are *double agents* who, by passing secrets to both sides, put in crisis the idea of an absolute Truth, in favour of a local knowledge that is measurable and achievable only within given strategic contexts. So, at the heart of the tale, we have a large number of Subjects: *pragmatic subjects* who undertake programmes of action; *cognitive subjects* who attempt to construct the other's being; *simulacra subjects* who are imagined by the other; *pretend subjects* who pretend to be something they are not.

In this way, passion appears forcefully, insinuating its way into strategic calculations, inevitably feeding and transforming them, determining their intensity and tension, duration and pace, desertions and new starts. The calculation of costs and benefits, risks and advantages, causes strategic calculations to lose track, allowing them to be transported by spontaneous gestures, by an 'irrationality' born of an excessive need for rationality. Though, as we will see, even passion possesses its own logic that feeds and enriches the narrative one.

III.6.4 Tactics

This is where *tactics* emerge: ulterior narrative programmes that are set in motion every time there is an obstacle to avoid. These are *counter-programmes*, in opposition to the programmes initiated by the adversary. Each action is not simply an effect of the one that precedes it and the cause of the one that follows. It is also the result of a tactical thought, a calculation of opportunities and risks, an evaluation of the other's real abilities. However, from a different perspective, given that the adversary carries out the same reasoning at the same time, the action is also a *signifying gesture*, a behaviour that will be interpreted: a *do* that is at the same time a *make-believe*. The opposite is also possible. Not only does every gesture signify, but every communicative process is a form of action, even when it involves texts that limit themselves to representing the ongoing conflict to the outside world. For example, it is well-known that information, during times of war, is always a strategic weapon. It participates in the *agon* by influencing the programmes of the forces on the battlefield, modifying their knowledge and passions, their skills. In the same way, a company that puts out a press release saying it wants to be floated on the stock market, or that pre-announces significant changes to its administrative board, is not merely providing its audience with information. Rather, it is constructing a particular image of itself and proposing it in the competitive arena. All communicative activity is a strategic move, a contribution to the continual, reciprocal positioning and repositioning.

III.6.5 Meta-strategies and culture

So, if doing is saying and saying is doing, it is necessary that the two contenders possess a kind of *meta-strategic ability* allowing them both to control and calculate the global strategic outcomes of local tactics. And it is indispensable for the same contenders to have something in common, a code with which to understand one another. If every contract presupposes a conflict (you agree to no longer speak to one another), then the opposite is also true: every conflict presupposes a contract (you must be aware of the shared stakes of the struggle). Polemical relationships and contractual relationships gradually and reciprocally blend into one another. If this weren't the case, we would fail to understand mechanisms such as the challenge to defend one's honour ("let's see if you're capable of . . . ") that are present in so many cultures, which imply a kind of obligation for the person challenged to submit themselves to the challenge or lose 'face'. Thus, the only way to avoid this is to leave the shared value system, to circumvent the blackmail of honour at all costs (the Gospel's advice to "turn the other cheek" is a perfect example of this.)

Strategy has its own culture of reference providing the Subjects' gestures, actions, behaviour, words and signs with different valorisations. Every culture has its own idea of the enemy, its ways to confront them, including that (more frequent than perhaps imagined) of not confronting them and preferring to steer clear of their moves without ever openly attacking. For example, Chinese culture is famous for its idea of development as a continual flow of tiny mutations rather than a series of discontinuous and predictable operations. For the Chinese warrior (but also for modern managers from the same culture), it is necessary to progressively adapt oneself, elastically, to the enemy – rather than fighting; to passively allow things to happen so that the adversary ends up exhausting themselves and the battle is won almost without any action. Similarly, certain ancient Arab warriors in the desert have a very particular strategic culture, made up of continuous routes of escape and sudden centripetal movements that confuse the enemy, leading them to defeat. So, before setting specific tactics and strategies in motion, we must concern ourselves with meta-strategically recognising these different cultures of strategy, understanding the *value of values* at play, identifying the codes of the other. This stands not only for large-scale cultural differences (which are pertinent at this time of globalisation and multiculturalism), but also for situations on a smaller scale.

III.7 Logics of affect

This raises the issue of affectivity. The 'way of being' of a Subject's identity or an atmosphere is the internal processing and articulation of states of conjunction and disjunction between Subject and Object. The narrative states, as we have hinted, have nothing static. In the same way, the so-called 'state of mind' is, in reality, an interior agitation of the psyche, real adventures of passion that do not stop or block action. Rather they are a profound motivational recharge that restarts the narrative programmes, giving new vigour to the value systems. Think of the strategy theory in which not only are there purely tactical passions (fear, courage, temerity, sense of risk, uncertainty and so on), but where the assessment of other people's skills, intersecting with the need for a demonstration of one's own, leads to an uncontrollable passionate 'outburst'. It is the same with the phenomenon of manipulation, in which every manoeuvre bound to the make-do and, in particular, the make-want, affects the pathemic dispositions of both the Subject to be manipulated and the Subject doing the manipulation. Not to mention the profound mechanism of the constitution of value, created by the projection of the timic category (euphoria/dysphoria), that belongs to the relationships of psycho-physical attraction and repulsion of a physical Subject with regards to their immediate environment. We must, therefore, redefine narrativity as a *guided process of transformation through actions and passions*, in which every action generates passion and, conversely, every passion provokes an action.

It follows that human and social meaning is produced not only on the basis of cognitive activity or codified cultural behaviour, but also through affectivity, which in turn makes continual reference to the somatic reality: the body. From this point of view, the first opposition to be overcome is that between reason and passion, as passion is the consequence of an action endured, but can also in turn generate it. Thus, the passionate dimension is not simply bound to desire and erotic affectivity, to physical sensuality and loving sublimation. Such a dimension is linguistically and culturally determined (there is nothing universal or intimate in passion). However, it possesses a profound layer that precedes its own linguistic and cultural organisation. In other words, beyond the ways in which languages name passions (often untranslatable between languages) and the forms that they acquire in various cultures, there are *passions without a name*: affective dispositions and modulations.

Semiotically, *passion is an effect of discourse*. The various discourses produce, solicit or transform the affects of the both Addresser and Addresser involved in the communication. Furthermore, affectivity is the final (and never definitive) result of a series of semiotic mechanisms, even beyond the passions expressly named or represented in the texts. If, as has been said, the *timic category* is the basis of every pathemic process, equally important are *modal combinations*. Jealousy, for example, has within it a wanting-to-know; stubbornness is a wanting-to-do that grows from a not-being-able-to- do; greed is a not-wanting-to-be disjoined from its own object of value; not to mention the fundamental passions of having-to, such as honour and revenge. Other phenomena that contribute to an effect of passionate meaning are *temporality* (nostalgia is a passion for the past, hope a passion for the future) and *aspectuality* (there are instant passions, such as horror, and lasting ones like anxiety, as well as inchoative ones, such as expectancy, and terminative passions like anger). Further phenomena contributing to the passionate dimension are *tension* (think of oppositions such as tense/laid back, stressed/relaxed, contracted/extended), *intensity* (which generates affective involvement of varying levels) and *rhythm* (a change in sound or colour acts as a conduit to clear passionate effects). These phenomena often intertwine, and affects come from and are modified by this intersection. We see this, for example, when a pain that is too intense produces momentarily euphoric reactions (like when we laugh at funerals), or when great tension transforms euphoria into dysphoria (when we cry out of happiness), or when the protraction of an affect is such that is causes distension and adiaphoria (when someone is jealous for too long to actually still feel it), or when we have sudden swings in temperament. Every passion is, in short, the result of these possible *montages* between different semiotic phenomena, and describing it means reviewing these montages and verifying, case by case, their reach. This entails an outcome that is never actually final since it is inserted into a process of continual transformations, timic rises and falls, tensing and relaxing, blockings and syncopes, renunciations and new starts.

III.8 Canonical schema of passions

The mechanism of passion is dynamic and process-driven. From this point of view, it is connected to (and surpasses) the articulations of narrativity in which every element acquires identity and meaning within a development guided by actions. As in the canonical narrative schema, the affective processes make reference to their own *canonical path*, which is made up of three fundamental stages (constitution, sensitization, moralisation), the second of which is, in turn, divided into three parts: disposition, pathemisation, emotion. We see this in the following schema:

constitution	sensitization			moralisation
	disposition	pathemisation	emotion	

Fig. 9: The Canonical Schema of Passions.

According to this model, every passionate phenomenon can be inserted into one of the five moments of a standard process of affectivity. This process is configured as a *crescendo*, which through simple 'temperaments' leads to an ethical catalogue of vices and virtues. We must understand a passion (or whatever any particular culture considers such) as an articulated series of stages each with a different semantic value. For example, anger is the result of a frustrated expectation of action by someone trusted, and for whom there is no desire for forgiveness. We can observe instances with affective value in any semiotic processes, even if they are not commonly recognisable or identifiable affects.

The first instance in the canonical schema of passions is the *constitution*. Here, the Subject's predisposition to access a passionate process is made manifest, starting with a *constituent* actant, depicted either as a real character or a series of external stimuli, be they social and family environments, historical and cultural circumstances, and so forth. For example, in the case of a passion such as greed, constitution consists of that kind of generic attachment to the things that a Subject could have acquired for a number of reasons from the environment in which they live and which has influenced them. Jealousy is that agitation caused by the seemingly strange behaviour of the beloved object. These are not real passions but pathemic propensons that, in certain conditions, can be culturally circumscribed quickly as affective states, or defined linguistically using passionate terms. They therefore belong to the realm of those passions we refer to as *passions without a name*, and in particular *without* (a single) *subject*, in which there is a kind of collective temperamental attitude.

The following three instances of the canonical schema of passions are grouped within the category of *sensitization*, where affective constitution becomes a true passion: a transformation created when a particular culture or language, a certain text or some imaginary universe interpret the affective square that has been previously constituted as a configuration of passions. To use our previous examples, an attachment to things is understood as specific greed; agitation in the face of the beloved object as jealousy. We are no longer faced with simple pathemic propensities, but recognisable and identifiable passions.

The first stage of sensitization is *disposition*, in which (as in the pragmatic phase of skill achievement), the Subject acquires the necessary abilities in order to prepare itself to be passionate in one way rather than another. Thus, the jealous person transforms their vague agitation into a *wanting-to know* that allows them to foretell precise forms of behaviour, while the greedy individual is able to organise their generic inclination as a *not-wanting-to-be disjoined* from their own goods, which may bring future actions.

The second stage of sensitization is *pathemisation*, an authentic passionate performance. There is no longer a passionate skill but an elementary scene with a series of recurrent themes and configurations, as when the greedy person imagines a transformation in the value of their own goods, setting in motion a programme for their defence. Or when the jealous individual believes they have been betrayed by their beloved Subject and obsessively tries to follow them, continually succumbing to theatrical and redundant scenes.

The third stage of sensitization is *emotion*, a consequence of passion on the Subject's body, the somatic manifestation of affect that tends to transform corporeity or cause it to act directly (blushing, stuttering, shakes, rash decisions). Whereas disposition has a cognitive nature, interacting with forms of knowledge and belief, and pathemisation has a pragmatic nature, as it gives rise to precise behaviour, emotion is based on the somatic dimension. The Subject's body becomes a vehicle for signification and communication. With *emotion*, the passionate process reaches the Subject's most profound intimacy (with its relatively autonomous somatic parts that act in an uncontrollable way) whilst at the same time opening itself up to the broadest sociality (exposing itself to public scorn).

The impassioned Subject's loss of themselves in an emotional moment leads to public outrage. From here comes the final moment of the canonical passionate path: the *moralisation*. It is with this stage that intersubjectivity comes back into play, as the passionate dispositives that were revealed during the passionate journey are brought within some kind of social rule (be it veridictive, aesthetic, religious, ideological, or otherwise), which tends to appear as an ethical norm. In *moralisation* there is an *evaluating* actant that operates according to the classic principle of *measure* (which is variable within space and time), according to which the *excess* or *insufficiency* of a particular passion is decided. It is in this stage that a passion becomes either a vice or a virtue. Returning to greed, it is thanks to moralisation that parsimonious behaviour is distinguished from meanness, with the first socially accepted, and the second, rejected.

A series of important clarifications are necessary with the canonical passionate path. It is first necessary to insist on the idea that, despite being faced with a schema (which like all others both explains and impoverishes the nature of the phenomena is wishes to explain), passion is a dynamic process that semiotics must grasp beyond any lexicalisations provided by various languages or beyond any recognisable configurations produced by various cultures. The aim of analysis is to defy linguistic and discursive stereotypes, explaining the procedures of construction. Secondly, the elaboration of the schema of passions is affected by the canonical narrative schema. As we can clearly see, constitution and moralisation (given also the presence of a specific actant that, respectively, sends and judges) refer to a manipulation and sanction, whilst the three moments of sensitization refer to the phases of competence and performance. This reference is more a virtue than a defect as it highlights the dynamic nature of passion (like that of the tale) and reiterates the necessary mixture of action and passion, of the pragmatic dimension and the passionate.

As such, the main characteristics of narrative logic can be repeated here in passionate logic. So it makes sense to distinguish between the discourse's actual actors and its actants in order to indicate their profound syntactical functions. The various actants of the canonical passionate schema (constituent actant, impassioned subject, evaluating actant) can be carried out by the same actor, or, conversely, from a number of actors playing the same actantial function. I can be my own moraliser ("Enough, I can't go on like this!"), just as I can have enormous ranks of constituents (relatives, friends, etc.). Furthermore, as this is a profound schema, not all instances on the path must be present on the text's surface, nor must they follow in linear way. The importance of this schema lies in the fact that it is enough to find just one instance of the path on the surface in order to be able to reconstruct the others deeper down, thus reconfiguring 'passionality' as an effect of meaning starting from just one of the schema's elements. In this way, it is significant to see which elements of the path are made manifest on the surface and which are, instead, hidden at a deeper level, so as to be able to reconstruct a dialectic of the implicit and the explicit in generative terms.

III.9 Forms of life

The progressive elaboration of narrativity as a general interpretative model of human and social experience begins with an almost exclusive attention to the meaning of actions. The problem has slowly opened up to the processes of the construction and transformation of identitarian subjectivity, to the interactions between cognitive strategies and inter-subjective tactics, as well as the sphere of passions and affects, in this way including at the same time corporeity and sociality. The cornerstones of narrative theory have been subject to constant integrations and revisions, in-depth analysis and expansions to the point that, as we come to the end of this chapter on narrativity, it is natural for us to ask: what is left of that *canonical narrative schema* that was posited as the theoretical key for explaining the formal mechanisms of the construction of human and social meaning? Now that the importance of the strategic and passionate dimensions are understood, does the schema seem to lose much of its heuristic value? If subjectivity is constituted and articulated through inter-subjective and affective processes, why conserve a model that seems to prize a rationalistic logic bound to the solitary achievement of a goal?

On closer inspection, the canonical schema focuses on the actions of a Subject charged with precise properties, on the calculations needed to reach the conjunction with the objects of value. Furthermore, the schema comes about thanks to the progressive generalisation of a precise narrative genre: the Russian fairy tale. This is a culturally, geographically and temporally circumscribed product in which the hero (the socially recognised individual) is the one who leaves to defend the values of the land and the kingdom, whilst the idiot is the one who prefers to stay safe in their hut, by the warmth of the fire, enjoying the most intimate affects of the family and the small group to which they know they profoundly belong. It is a very precise cultural conception that tends to valorise action over passion, institution over sentiment, doing over being. It is a conception that though seemingly repeated and amplified by certain mass media products in some respects, can seem very partial when considered in light of the multiple, complex, geopolitical and socio-cultural transformations of today's society and world in general (post-colonial processes, the debate regarding gender, globalisation and localisation, multi-culturalism and so on).

The canonical narrative schema is a model that, though it should not be abandoned, does need to be scaled down, starting by eliminating that aura of universality sometimes attributed to it by superficial interpretation or mechanical use. It is a good starting point for explaining certain mechanisms for the production of meaning, but it is not the only possibility. To this model we must progressively add a series of others that recognise more complex and more refined socio-semiotic phenomena than the simple succession of its four standard stages (manipulation, competence, performance and sanction) is capable of explaining. It often happens that a Subject's lifestyle (individual or collective) is constituted through its difference to the standard narrative model, that the Subject doesn't structure precise narrative programmes, be they base or instrumental, doesn't valorise specific objects by making them their existential objective, is not concerned with acquiring skills or interested in social realisation. This Subject does, however, allow themselves to pursue ethical or aesthetic goals that are hard to reconcile with cultural macro-organisations. To projective subjectivities based on cognitively controlled decisions we can add other forms of subjectivity and experience that are more interested in affectivity or aesthetic expression, or in the care and exhibition of the body. These are *forms of life* with significant differences between them but that can all probably be traced back to a *coherent deformation of standard models of civil living*, of those strong social codes that, once introjected individually, appear obvious and natural. A deformation that, once it has occurred, can be more or less established, enter into common use, act as a system and become a source for negotiation in the social arena.

An example: a young maiden throws her glove into the lion's den, promising her love to the knight who goes to collect it. A knight accepts her challenge, faces the task, returns the glove, but refuses her love, despite having longed after her for a long time; then he leaves, alone. The knight's unexpected gesture is an absolute negation of the narrative logic that dictates a Subject submit to a task in order to conjugate with their own Object of value. What is valorised here is the aesthetic of behaviour, action deprived of utility and, for this reason, equipped with a substantial ethical importance: maidens should not give themselves away, and real knights should not win them because of their ability in battle. So, the entire system of values ends up being overturned, and the anthropological mechanism of social reciprocity based on the nexus of gift/counter-gift vanishes. How many advertising spots have this kind of narrative construction, in which the irony that comes from breaking with traditional narrative schemas leads to a resemanticization of human and social experience?

The apparatus of narrative semiotics takes on (as well as their canonical schemas) their socio-cultural and even individual variations, demonstrating the close nexus between, for example, the ways in which a product or service is consumed, and the corresponding forms of life assumed and exhibited by the subjects involved. So, for example, drinking a coffee should not be understood as the simple act of ingesting food. Such an action recalls an entire way of understanding existence, of organising time and its correlating dramatization. Some drink coffee as an accompaniment to various activities throughout the day, causing it to interact with those activities, while some, conversely, use it to interrupt those activities, a kind of restorative break in which appreciation of the drink demands an isolation from the social context. There are also those who gulp their coffee down, sacrificing the delights of the palate to the pressing rhythms of life, and those who, on the contrary, abandon themselves to savouring a particular blend in an occasional or even casual way.

The exhibition is fundamental for the forms of life. The terrorist carries out actions that become significant not for what they are but for what they signify, only if they are picked up and publicised by the media. In the same way, the dandy is an individual who shows a certain contempt for the people and the world around him in order to provocatively draw attention to himself, slavishly playing his own unlikeable part for those very people he looks down on. On the one hand, the theatrical dimension connects ethics and aesthetics. Called upon by an unexpected gesture, the enthralled spectator rethinks their own values, comparing them with those of the other and ends up reformulating their shared common worth ("Is that *really* how things are?"). On the other hand, the gesture represents itself as a part of a whole. A text, more than being a simple fragment of a much bigger culture, can effectively express the entire culture, condensing it within itself. In the same way, a small gesture, the detail of an action, the aesthetic form of a behaviour can manifest the entire form of life of someone who, by assuming it, exhibits it publicly. A way of dressing is a way of life, in a profound tension between contraction and expansion, manifestation and depth, textuality and narration, going well beyond the world of clothing to invest individual and collective life.

The semiotic procedure on which the form of life is based is *the dialectical alternation between contraction and expansion*, between locally expressed figures and global configurations that subsume one another causing them to signify, between small gestures full of potential expressivity and systems of meaning that allow for its semantic realisation. In order to be a form of life it is necessary for a Subject to select a semantic category ('perfection', 'incompleteness', 'iteration', 'vagueness', etc.) and make it *dominant* within their own existential organisation, as if all of the other categories have to derive from it or lead to it in some way, in a continual game of expansion and contraction. So, if the classical artist's form of life descends from the category of completeness, that of the existentialist derives from incompleteness. While the iteration of risk is positively embraced by the drug addict as a function of the pleasure they seek, dominant for the romantic hero is the idea of destiny (with the conclusion of their acts and the success of their behaviour always emerging in their form of life). Similarly, looking at an entirely different semantic field, there are those who select the category of choice, placing it within an existential regime of *distinction* with respect to others and the world, and those who, conversely, choose the category of 'mixing', preferring the regime of *fusion*, realising themselves in an identitarian way by dissolving into a group.

Recommended bibliography

On the semiotic square: Greimas & Rastier 1968; entry "Square (semiotic)" in Greimas & Courtés 1982.
On narrative and narrativity see: Adam 1984; Barthes 1975a, 1975b; Barthes (ed.) 1966; Brandt 1992 (on modalities); Bremond 1973; Calame 2000, 2019; Calloud 1976; Cobley 2014; Dorra, Filinich, Moreno, Rodríguez Vázquez & Zepeda 2019; Eco 1995; Flores 2017; Greimas 1983a, 1987c, 1988 (analysis of a tale of Maupassant), 1989; Greimas & Courtés 1982, 1986; Lozano & Salerno (eds.) 2020; Lorusso, Paolucci & Violi (eds.) 2012; Pavel 2017; Perron 2003; Ricoeur 1984–1988.
On strategies and interaction: Alonso 2005; Calame 2019; Ducard 2017; Fabbri 2000; Fontanille 2017; Goffman 1967, 1969, 1971; Jullien 1986; Landowski 1989, 2005; Mitropoulou 2017.
On passions: Fontanille 1993 (on the canonical schema of passions); Fabbri 2001; Fabbri & Pezzini (eds.) 1987; Greimas & Fontanille 1993a; Landowski 2004; Barros 2017; Moreno 2014; Parret 1986; Pezzini 1998; Pezzini (ed.) 1991; Pezzini & Del Marco (eds.) 2012; Rastier 2019.
On forms of life: Basso, Bertrand & Zinna (eds.) 2018; Fontanille 2015; Greimas & Fontanille 1993b; Marrone & Mazzucchelli (eds.) 2019; Moreno 2014; Zilberberg 2011.

IV Enunciation and discourse

IV.1 From communication to enunciation

The study of narrativity has provided a series of models that allow us to describe the profound articulations of its textual content. It is now time to shift our gaze to a new subject, that does not refer to the content or values enunciated in the text but the way in which they are enunciated, i.e. the communicative strategies enacted in order to transmit that content as well as the interactions between the instances of production and reception. In short, the *discourse*, which must be examined in all of its importance and complexity. If narrative is the most profound form of communication, who is telling these stories? What kind of narrator are they? Do they openly reveal themselves or prefer to remain hidden? Do they take responsibility for what they say through the use of the first person, or do they delegate to more authoritative figures? Do they know what they are recounting or are they simply repeating other people's words? Furthermore, to whom are these stories told? To someone who is curious and competent, or a subject whose interest must be captured, first, and then convinced and educated? And what is the relationship between these two subjects? Agreement, conflict, collaboration, negotiation, friction? The question is, therefore, how are these *pacts of trust,* without which no communication can take place or have meaning, constructed and managed? And what is the nexus installed between the kind of relationship created between an enunciating subject and the general communicative efficacy of their discourse? How can all of this be analysed and reconstructed *a posteriori*?

We are at the heart of communicative processes. They must be understood not in the sense of empirical production and reception by tangible actors of the communication but, rather, examined as an inscription of the production, the reception and their actors within the flow of discourse, and therefore in the texts that such discourses tend to manifest. In technical terms, this means reinterpreting communication as *enunciation*. From this comes a more relaxed and more circumspect attitude to communicative phenomena. On the one hand, *communication* is inserted within the broadest framework of *signification* that considers not only those texts constructed with explicit communicative aims (a book, a film, a painting, a lesson . . .), but also those which would not have this prerogative, and that do not therefore present themselves in this form. For example: the articulation of spaces in a city, the design of everyday objects, the micro-actions carried out by a lift, the routes taken by commuters on the underground, the rituals of a popular festival, interactions on social networks. On the other hand, the semiotic gaze tends to play down the concrete communicative contexts (casual or constructed) of production and reception, together with the economic motivations (political, social, familial, emotional, etc.), that lead a given social actor to propose a particular *communicative pact* to a given public that accepts (or refuses) it. Semiotics rediscovers these circumstances inscribed in the discourse, according to the basic semiotic principle by which the *communicative context of a text is inevitably present*.

The starting point for understanding the semiotic notion of enunciation is an analogy: every product presupposes a producer, i.e. every human artefact exists because a creator existed previously, one who thought of it, planned it for some purpose and consequently built it; in the same way every communicative product presupposes someone who has communicated it (who has wanted it, thought of it and issued it for some reason), as well as someone to whom it is directed. No matter the substance or the form it presupposes, an utterance acquires most of its meaning not from the external world to which it refers, but from the person who has made it and the person for whom it was made. In other words: every message emitted, in any communicative process, presents within it not only a particular enunciated content, but also a trace of the process of its production, a kind of secret signature of its author, a silent appeal to its end-user. Consequently, every enunciation has, by the very fact of its existence, its own 'mark', no matter its thematic content. This is the sign of the act that has brought it into being, of the subject who made it, of the other subject to whom it is directed, and of the practical consequences of all this. The *discourse* is a phenomenon that contains the communicative process, its product, the actors that produce it, the marks left by the process and the actors on that same product.

Such an issue becomes, at the same time, more evident and more extreme if we consider not only the intentionally emitted processes of communication, but any significant fact or event that happens in the world regardless of any direct, explicit, or conscious communicative will. Not only does every communicative product presuppose a structure (explicit or implicit) that frames it, providing it with further significance. More generally, however, every action equipped with meaning, intentional or otherwise, exists as an enunciative instance thanks to which it can offer itself to the social discourse and manifest itself textually. Any signifying praxis has within it the traces of production and reception that indicate the source and the target, influencing them. This is the case whether the signifying is part of an advertising campaign, television programming, an internet portal, a fashion show, a shopping centre, a city, a collection of photographs, a politician, or an Instagram post.

IV.2 Languages and subjectivity

The linguist Ferdinand de Saussure (1988) distinguished between two aspects of language: the *parole*, i.e. the moment in which the individual actually communicates, and the *langue*, i.e. the formal system of rules shared by speakers. There is a variable aspect of language by which every speaker expresses themselves in a different way, and an invariable one, an abstract and social code that, transcending the will of the individual, guarantees the success of their communication. *Parole* and *langue* exist in a dialectical relationship: while the *parole* requires the *langue* in order to manifest itself, the *langue* only exists as the final result of the individual ways of speaking. This dichotomy, along with its numerous revisions (code/message, competence/performance, profound structures/superficial structures, etc.) forms the basis for the first semiotic research in a structuralist framework, which pursued the aim of searching for invariants beneath the variables, the forms that articulate the substances, the systems that allow the processes to take place. Any form of subjectivity and historicity is discarded as an occasional manifestation of underlying codes, a variable of little importance to the constitution of a super-historical and tendentially universal system that makes the apparition of meaning possible. From here comes a static and authoritarian image of semiotic systems in which the individual seems to have little freedom of expression and action, a pawn that unwittingly repeats what is dictated in the social codes to which they submit. And from here comes a series of criticisms of this restrictive vision of language in the name of inventive capacity and the creative acts of single speakers (just think of literary writing). Some have affirmed, not without provocation, that *langue* is, as such, fascist.

IV.2.1 Enunciation in philosophy

Numerous scholars have indicated a number of paths for overcoming this difficulty. Ludwig Wittgenstein insists on the importance of the effective use of language in order to determine the meaning of linguistic terms. Charles Morris (1938) uses the term 'pragmatic' to indicate that specific part of semiotics that deals with all the psychological, biological and sociological phenomena that intervene in the functioning of signs. John Austin points to a series of linguistic facts, generally neglected in studies of language, by which the subjects implicated in communication play a primary role. Next to the constative utterances whose aim is to describe the world, there exists in language a dense series of utterances that, according to Austin (1962), are also forms of action. If "the Earth is round" predicts something about the world, utterances like "I declare this session open", "I'm sorry" or "I welcome you" say nothing *about* the world, but do something directly, they act *in* the world, transforming it. Constative utterances, taking on the burden of representing reality, are subject to the criteria of true or false. A second kind of utterances, referred to as *performative* by Austin, are instead neither true nor false, but rather happy or unhappy. Their value does not lie in their descriptive ability, but in the appropriateness, in the fact they are said at the right time by the right people. In this way, single acts of linguistic utterance are no longer simple manifestations of an underlying code, as widely-accepted Saussurien thought would have it, but moments in which the rule system that allows communication to succeed or fail is constructed or validated.

IV.2.2 Enunciation in linguistics

In parallel with the philosophical reflection on the pragmatic uses of language, linguistics has progressively dealt with the mechanisms that inscribe the speaker and the listener into the language they use. The theory of communication functions proposed by Roman Jakobson (1963) indicates how the possible ways of using language (emotional, phatic, conative, and so on) are inserted in language in the form of grammatical and syntactical rules that codify their usage. A mode of the verb such as the imperative, or a case like the vocative, are specifically linguistic places through which the conative function is executed. In the same way, interjections are that part of the discourse charged with grammaticalizing the emotive function of communication centred on the source. Jakobson's functions are not actual uses of language but the way in which these uses are provided for in advance by the linguistic system.

The scholar who insisted on the formal presence of the speaking subject within verbal mechanisms is Emile Benveniste (1971, 2015). This linguist has repeatedly asserted that a whole series of categories (personal pronouns, demonstratives, verbal forms, modes of verbs) can only be explained by considering the particular situation of enunciation in which it is being used. As such, if common names are endowed with a codified signified in a language as they refer to established objects, the pronouns change signified every time depending on who enunciates them. For example, "I" is the speaking subject, and its semantic value is modified each time the person speaking changes. In the same way, "you" refers to the listener, whilst "he" is someone who is neither the speaker nor the listener. The same thing happens with the present, the past and the future, temporal categories that take on meaning only in relation to the act of speaking; or with demonstratives, whose effective meaning varies depending on where the speaker is. Beyond the systems of rules that exist within language, and beyond the single linguistic utterances, there is a formal apparatus of rules that, demanding the tangible presence of speakers, refers to the linguistic code. *Enunciation* is the mediation between *langue* and *parole*, which is manifested in the tangible communicative act and that is provided for in advance by language, at the same time.

The formal apparatus of enunciation allows subjectivity to emerge and be constituted through language. Not only is the single subject made through 'empty forms' that language places at their disposal, but inter-subjective relationships also depend on the way in which the situations of discourse resort to the linguistic codes that allow for it. As such, if I say "I order you to open a window", I am proposing myself as someone who is allowed to give an order, and I construct my receiver as someone who should take one: I create a social hierarchy. While Austin insists that the conditions for communication are the measure of the happiness or unhappiness of a linguistic act, Benveniste believes it is the linguistic utterance that produces the correct conditions for enunciation to be effective.

IV.2.3 Enunciation in semiotics

Enunciation is not only a linguistic phenomenon, but a more generally semiotic one. In literature, for example, the problem of the narrative voice and perspective (ways in which the author presents themselves or hides within the work) is similar to that of linguistics in enunciation. In the same way, painting, through a complex system of gazes and gestures, has its own specific ways of saying "I" and "you", inscribing the painter and the spectator within the text. Take also cinema, which constructs real 'audio-visual conversations' through its specific techniques, such as subjective shots, looking directly at the camera, the *mise-en-scène* of the set and so on (Casetti 1999). Not to mention television, where auto-referential tendencies can be observed with the demonstration of itself as an enunciative apparatus (Casetti 2002). Even a city, with the organization of space, constructs an image of itself, its own history and the values it intends to propose to those who, by inhabiting it, tend to create an idea of it (Shapiro 1973). Hence the need for a semiotics that unifies these different questions, taking a series of observations from philosophy and linguistics in order to put them to work in the analysis of any text, be it verbal, visual, gestural, spatial or otherwise.

Semiotically, every utterance, no matter the substance of expression to which it resorts, presupposes an enunciation, an original productive act that can then be manifested to different degrees within the enunciated itself. There can be cases in which the subject of enunciation is explicitly signalled (with a first person pronoun, the movement of a film camera, the depiction of the artist on canvas, etc.), or those in which every trace of enunciative production is hidden (the linguistic "they", figures painted in profile, the absence of authorial intrusions in literature, etc.) so that the enunciated text appears as if suspended in the void, deprived of any reference to the person who produced it and, therefore, projected entirely towards the 'reality' that it tends to represent. The enunciation is always present in the utterance, even when it is imperceptible, as the absence of its being made (signalling the desire to construct forms of 'realism') appears more significant than its presence.

It is possible, therefore, while analysing the enunciated text, not only to reconstruct the semantic structures of the message, but also the enunciative structures that have created it. We just need to consider how the enunciated is the result of a primary implicit foundational act called *shifting-out*, thanks to which the three fundamental categories of actor, time and space come into play, effectively passing from the level of narrativity to that of discursivity. If the subject of the enunciation is an "I" that speaks in a "here" and a "now", the utterance could reproduce these figures within it (essentially enunciating the enunciation through a shifting-out that is *enunciative*), or it could deny them, founding itself on a "not-I", a "not-now" and a "not-here" (and therefore cancelling the enunciation with shifting-out that is *utterative*). The opening words of fairy tales – "once upon a time, in a faraway land, a king . . . " – is the most typical result of an utterative shifting-out. Certain forms of autobiography that recount to the present the "I" that is writing are an example of enunciative shifting-out. Once the categories of actor, time and space have been installed within the enunciated they can be modified, providing innumerable cases of further shifting-out, or cases of shifting-in, a return to previous categories. So, for example, the classic 'tale within a tale', used by much realist literature, is a case of shifting-out within the enunciated that sets in motion a second story, which is generally followed by a return to the initial narration (a shifting-in). We see something similar with news programmes in which the television broadcaster hands over to the presenter (each time modifying space and time) who, in turn, hands over to the correspondent, who often then hands over to the person being interviewed, in a series of progressive shifting-out and consequent returns, or shifting-in, towards the initial enunciative instance.

IV.2.4 Enunciation and action

If enunciation is, in principle, an instance presupposed by the utterance, it is because this can be interpreted as a form of action. Just as any object is inscribed with the traces of their producer, an utterance, in the same way, holds within it the *marks* (I/not-I, now/not-now, here/not-here) that refer to the subjects of the enunciation, so, to the *Enunciator* (a textual simulacrum of the person who produced it) on the one hand and, on the other, the *Enunciatee* (a textual simulacrum of the person it is addressing). Where Austin distinguished between constative and performative utterances, for semiotics any linguistic enunciated is a semiotic act, whether performative or constative. Even ascertaining a state of the world is a form of action that has precise effects on the enunciatee and presupposes certain given intentions for the enunciator. The constative is that kind of utterance that hides its own enunciation in an enunciative shifting-out. Thus, for example, an enunciated like "the Earth is round" is a product that has previously cancelled out the shifting-out needed to construct it, so something like "(I, here and now, say that) the Earth is round".

In this way, if the enunciation is an action, it can be studied using *narrative models*. Communication becomes a tale, and its fundamental characters (source, message, receiver) can likewise be thought of as narrative actants. Speaking is not just transmitting knowledge, but setting in motion an action in which a Subject of Doing (Enunciator) conjoins a Subject of Being (Enunciatee) with an Object (the message). And, as we know, it doesn't matter so much what this Object is so much as the value inscribed within it. Thus, the Enunciator is not only a Subject of Doing but also the manipulating Addresser that inscribes the value of 'truth' (or 'beauty', 'justice', 'propriety', and so on) onto the Object-message, proposing it to the Enunciatee. The latter is not simply a Subject of Being who passively submits to their conjunction or disjunction with the Object-message, but also a Addresser-judge who judges the value of 'truth' inscribed within the Object-message, accepting or rejecting it.

This has a number of consequences. Firstly, indicating the protagonists of the enunciation as *actants* means re-proposing their difference to the other effective *actors* of the communication. The Enunciator and the Enunciatee are not real people who emit and receive the message, but their textual simulacra, and as such, they can be actorialised to varying degrees. This means that if the communication is an exchange, the roles of the Enunciator and the Enunciatee alternate between actors who remain physically the same over the course of a conversation. But what is most interesting, is that between actants and actors there is almost never a one-to-one correspondence. Thus, for example, an entire television network, with its apparatus of people and technologies, assumes the role of Enunciator (unique from a textual point of view) beyond the effective number of presenters on the scene; and the entire television audience is, generally, an Enunciatee, also textually unique, whether they are absent from the place of broadcast or represented directly, as often happens with a 'live studio audience'. Furthermore, the narrative roles of the Subject on Doing and the manipulating Addresser (on the side of the Enunciator) and the Subject of Being and Addresser-judge (on the side of the Enunciatee) can be embodied by a single actor (like in some news programmes where the presenter is also the carrier of the values in play), or by many (as we see in those British newspapers that tend to distinguish clearly between news and comment).

Secondly, as actants, Enunciator and Enunciatee are variously charged with modal values (to want, to have-to, to know-how, to be-able-to) within the text. Whereas in traditional communication theory the source and the receiver are instances of transmission and reception with no particular internal determination, in semiotics the Enunciator and the Enunciatee are differently constructed forms of subjectivity that enter into a relationship through different modal charges. So, it's one thing to speak to an audience that is endowed with having-to (as in a didactic discourse), and quite another to speak to a receiver that is endowed with wanting (as in a journalistic discourse). Likewise, it's one thing to speak to someone presenting oneself as a powerful subject, it's another to speak to someone when this is not the case. The possibility of combinations is, as we can clearly see, very broad.

Thirdly, the idea that the Enunciator and the Enunciatee, as well as being pragmatic subjects (Subject of Doing, and Subject of Being) are also cognitive subjects (manipulating-Addresser and Addresser-judge) introduces to the communicative exchange, before the action, the moment of the *contract* – a more or less tacit agreement on the values that will come into play during the tale, on the respective roles of the subjects involved, and their eventual hierarchy. This contract can, depending on the era, culture or communicative situation, be presupposed by the text itself or stipulated progressively as the text develops, or it can be further transformed depending on the communicative strategies to be set in motion. The result is an idea of communication and language that is very different from the traditional one. The criterion of an utterances's truth or falsity is not so much determined by its relationship of adequacy to the external reality (referential relationship), but from the relationship between Enunciator and Enunciatee, that, on the basis of respective modal charges, can find an agreement on the truth of the communicative process (communicative relationship). The truth, from this perspective, is not the effect of a representation but the result of an inter-subjective relationship.

IV.3 Efficiency and efficacy

As we have said: we must not confuse the source and the receiver of standard communication theory with the Enunciator and the Enunciatee of the semiotics of enunciation. While the former are empirical actors, tangible entities (or technological apparatuses), the second are their simulacra within the discourse: their semiotic 'images' that, depending on the textual substance used and the desired communicative objectives, can take on the most varied forms. They are simulacra that the analysis identify and reconstruct with the aim of understanding the kind of communicative pact of the discursive process.

Enunciator and Enunciatee are not representative of real communicative actors. In order to better understand the real function of the discourse as a phenomenon that is semiotic and social, empirical and tangible, we should ask ourselves why the discourse uses these two figures. Not only, as it might seem, in order to remember the moments of its production and reception but, more so, in order to provoke them, to bring them into being, to make the enunciative process a communicative strategy. Strategy is a narrative notion. As we know, in strategies (be they for war or otherwise) the simulacra of the Subjects at play are forces in the field: in order to be able to fight them, or reach an agreement, the Subject and Anti-Subject each provide an image of themselves to show the other while, at the same time, each constructing an image of the other so that these simulacra are real weapons to be used against the enemy. Similarly, in communicative strategies, in order to interact, source and receiver must first reach an agreement on the values of the communication, negotiating and fighting, fighting in order to negotiate. They therefore use simulacra of themselves and the other as persuasive weapons against the other. The source proposes an image of themselves (the Enunciator) and an image of their receiver (the Enunciatee). The receiver does the same thing, constructing an image of the source (the Enunciator) in the same moment they provide an image of themselves (the Enunciatee). The communicative pact resulting from this double strategic image of the communicative actors within the utterance affects the final communicative efficacy. In other words: the tangible *efficacy* of communication (be it cognitive, passionate, pragmatic or somatic) derives from the internal discursive *efficiency*, thus from the interactions between the communicative actors in the structure of enunciation.

Therefore it is the Enunciator and the Enunciatee that tangibly carry out the communicative action. Whereas the source and receiver are simply terminals of such actions, abstract instances posited as the starting and finishing poles of a process, Enunciator and Enunciatee construct the meaning of the utterance, providing it with values, and proposing it as such in the social arena. Saying 'I' does not mean exhibiting an image of oneself but, as Benveniste said, constructing oneself as a communicative subject. In parallel with this, saying 'you' means involving someone in the discourse, coveting them. Enunciator and Enunciatee are not representatives of the acts of communication but their prodromes, a kind of *instructions for using the discourse* inserted within the texts that manifest it. They tell us the textual genre in which the utterance should be inserted; they provide the necessary passionate load to use it, and the system of values necessary for appreciating and judging it; they outline the boundaries of behaviour that this might cause; and they prefigure the somatic transformations that this provokes. They do this without, however, presenting a one-way communicative situation in which the receiver necessarily finds themselves coinciding with their textual image, the Enunciatee. The Enunciatee should, in fact, be understood as a *proposal of meaning* that, in turn, the receiver can either more or less accept or more or less refuse: adhering sometimes partially, sometimes entirely, with the figure of the enunciatee but, most importantly, constructing in turn an image of the source as enunciator. All communication is a form of conversation. Identification, rejection, adhesion, distancing are all possible forms taken by the Enunciatee's response to the Enunciator, and the latter may have previously allowed for each of them.

The virtuous cycle set up between enunciative pacts and communicative efficacy becomes evident. On the one hand, the stipulation of a pact is a condition for the success of the communicative/ significative process, so that trust in the signs and their producer is a constitutive precondition of every semiotic act. On the other hand, the trust that the Enunciatee places in the Enunciator and the consequent communicative pact between these two actants is not preliminary. There is no initial moment of contract that gives rise to the discourse. It is in the process of the communication that such a pact is stipulated, thanks to the many possible ways in which the Enunciator and the Enunciatee reciprocally construct one another within the discursive flow, maintaining polemical-contractual relationships that are more or less complex on the basis of their respective modal charges. A communicative pact is not stipulated: rather it is *re*-stipulated, recharged with meaning or provided with other values depending on the sociocommunicative situations in which it finds itself.

IV.4 Strategies of knowledge

Beyond the procedures involved in constructing the discourse's veracity, explained by transposing the narrative logic at the level of enunciation, the problem of this same discourse's content remains the knowledge exchanged between Enunciator and Enunciatee. The utterance, as we have seen, can be interpreted as an Object of value that the Enunciator transfers to the Enunciatee. The fact remains that the utterance is also an informative configuration that contains knowledge. A series of questions emerges here. For example: does the Enunciator always know what they are talking about? Do they know what they are saying, or do they delegate responsibility for their discourse's content to someone else, a tangible character or abstract instance? In the same way, to what point does the Enunciatee appropriate the knowledge circulating in the discourse? Within the discourse, how is the news constructed, how is it shared with the reader, is it given as a certainty or an uncertainty? Any text tends to construct a complex cognitive configuration that we call "news" or "theoretical hypothesis" or "message", providing them with particular characteristics and offering them to the Enunciatee.

This is how a series of procedures specific to discourse are established, aimed at constructing the transmitted knowledge or, to borrow a formula from literary theory, to *regulate the flow of textual information* (Genette 1983). The literary notion of *point of view*, redefined in semiotic terms, becomes a general interpretative category with which to analyse the cognitive dimension not only of literary texts, but of cinematic, pictorial, spatial, journalistic, scientific and any texts of other kind. The subject of enunciation, as well as being a pragmatic actant or someone that *does*, having to exchange an Object that is a message, must in principle contain within it (and at the same time distinguish itself from) a cognitive actant (someone that *knows*). And just as an Enunciator and an Enunciatee exist, giving rise to communicative performances, it is possible to hypothesize the existence of that which, within the discourse, we can call *enunciative subjects*, an Informer and an Observer that exchange specific forms of knowledge. The Informer is a subject that offers themselves in varying degrees to the Observer's cognitive capacity, and the latter is, vice versa, a subject that is endowed with a cognitive skill. In this way, contractual regimes (for example: one wants to make themselves known, the other wants to know them) or polemical ones (for example: one does not want to make themselves known whilst the other wants to know them) can be instituted between Informer and Observer. And knowledge is the result of these cognitive exchanges, which can be interpreted once more using the model of narrativity.

Observer and Informer are actants and can therefore be manifested in the form of actors or remain abstract and implicit instances that analysis must work to clarify. In painting, for example, there is an implicit Observer who is easily identifiable in pictures that use a classical perspective. Here the point of view with which the image is constructed coincides with that of the empirical spectator. There is also an explicit Observer present in the painting when one of the depicted characters looks towards the painting's main subjects or scene. In literary descriptions, there is an explicit Observer when they talk of someone who, looking out a window or from some other observation point, looks at the landscape being described. There is an implicit Observer when it becomes possible to reconstruct the path of a gaze that moves from high to low, or from left to right, through the description. But the categories of explicit and implicit are not sufficient for identifying the observer. Indeed, depending on the case, they can remain entirely indeterminate (*focaliser*), they can receive a spatial and temporal placement (*spectator*), they can be expressed through an actor (*assistant*), they can become a character in a story (*participating assistant*), they can be the central figure (*assistant-protagonist*). So, for example, in television, the Observer can remain hidden when the camera frames the studio in a traditional way without making any particular movements, or making its presence known. Equally, they can place themselves within a precise space where the movements of cameras can be perceived by spectators. Sometimes both in the studio and in external locations cameras themselves are framed so that they act delegates of the Observer. Other times, there are human figures who observe: the presenter who turns to face the screen and takes part in the piece along with the audience, the correspondent "on the ground" observing the scene, a testimonial figure who was present when the events took place, and so on.

If the Observer is an intermediary actant between enunciation and the utterance, who *knows that that there is something to know*, the Informer is a second intermediary actant who *knows that there is some knowledge to impart*. As an actant, the Informer can also be figuratively expressed through an actor (take the classical messenger of Ancient tragedies) or remain partially or entirely implicit (think of the so-called "undisclosed sources" of politicians or journalists). The famous *commentator* referred to by Leon Battista Alberti – the figure present in the scene depicted in a painting who indicates what must be looked and must therefore be considered important – is a typical 'actorialisation' of the Informer. We must, however, be careful to avoid errors in terminology: the Informer should not be identified in a strict sense with the person who informs (in this case they would simply be assimilated with the Enunciator and its delegates sometimes present in the utterance), but rather with the possible discursive figures that provide, to varying degrees of satisfaction, the necessary skill to do so. In scientific texts, for example, the tools used in experiments (microscopes, computer screens, test tubes, etc) act as Informers that try to counteract the eventual *resistance* of the object to making itself known. The Informer *par excellence* is the spy, the informant, the confidant, the "undisclosed source", the expert, all actors present in the case where this cognitive actant is endowed with positive skills (wanting, being-able-to, etc.) that facilitate the circulation of knowledge. But this is not always the case. Often the Informer's skills are negative (not-wanting-to, not-being-able-to, not-knowing-how), leading to the emergence of figures such as the double agent, the unreliable source, the official bulletin that restrict the channelization of knowledge to the point of impeding it entirely.

Endowed with variable modal skills that are positive or negative, the Observer and the Informer can maintain relationships between them that, much like those of narrative actants, are at times contractual and at others polemical. The lucky case is that in which the Observer wants to know and the Informer wants to impart knowledge (in the way newspapers use to present themselves), but it can also happen that the Observer wants to know, but the Informer does not want or is not able to share knowledge (such as with a governmental meeting held behind closed doors, media blackout, secret negotiations by military forces, etc.). It can also happen that the Observer does not want to know and the Informer wants to share knowledge, thus forcing the first to acquire information they do not desire (this is frequently the case with scientific texts, which tell of an unexpectedly emerging reality that demands to be taken into consideration).

Furthermore, the regimes of knowledge are configured differently depending on the multiplications and the relative unifications of both the Observer and the Informer. It is clear, for instance, that if there are many contradictory information sources, the knowledge tends to crumble, to lose credibility, whilst if there is just one Informer, a kind of realistic, truthful discourse is produced. In the same way, if there are many observers the truth tends to fragment, whilst if there is only one, a kind of subjective certainty is created. The result is the following schema, which illustrates the various possible cases:

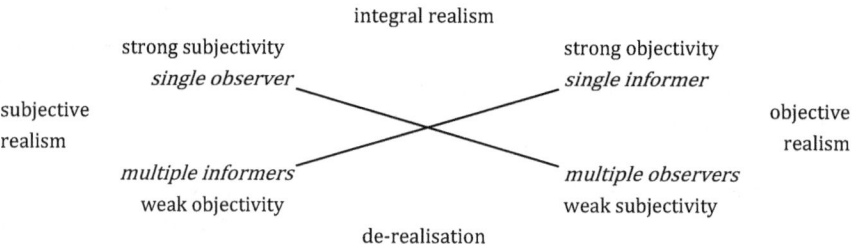

Fig. 10: The semiotic square of the regimes of knowledge.

A cognitive structure typical of political information in crisis situations or moments of institutional change is that of subjective realism, in which the observer-journalist is a single agent who attempts to bring together a complex, varied framework, from which emanates a multiplicity of possible informers, sources, voices, rumours and denials that weaken the objectivity of knowledge. The result is an explicitly subjective point of view, in which the entire truth of the discourse is delegated to the skill of the Observer and, often, of the enunciator who takes on to themselves (through shifting-out) the functions of this cognitive actant ("Now I'll tell you how things really are"). But there is no guarantee things will always go this way. Often the multiplicity of Informers and Observers produces a de-realised cognitive configuration that does not allow Subjects of state to conjoin with a certain knowledge. Only after having established a contract of strong veridiction between Enunciator and Enunciatee can the Enunciator take the function of Observer, assuming first person responsibility for the knowledge and demonstrating who the real Informer is. So, from de-realisation, we move to its opposite, integral realism, in which both the Informer and the Observer are singular. The credibility of the discourse is ensured.

IV.5 Intertextuality, interdiscursivity, intermediality

A further issue regarding the phenomenon of enunciation is the way in which the discourse empirically manifests itself. We have seen that the notion of discourse and, with it, the Enunciator and Enunciatee actants, despite generally having clear efficacy and communicative results (pragmatic, cognitive, passionate and somatic) that are entirely real on a social and cultural level, have an abstract existence and a simulacral nature. As semantic entities, their capacity and strategic strength derive from the fact they can only be perceived thanks to mediation by some textual substance, regardless of whether it is intentionally produced and knowingly perceived or otherwise. Only the text presents the double face – expression and content – that constitutes every language. On one hand, it is the semantic result of the utterance and the enunciation, thus producing a discourse. On the other hand, to do this, the text needs to articulate and form one or more physical substances that must express that semantic result.

Thus, the *discourse* can be manifested by very diverse *texts* in a mix that includes both communicative tools generally recognised as such (books, images, videos, songs, cooking recipes, etc.) and entities generally considered to be non-semiotic: physical spaces, behaviours, strategic actions, consumer experiences, but also objects and their designs, instructions for use, user interfaces, forms of inter-objectivity. This relationship between abstract discourse and the empirical texts that manifest it is reciprocal: there is no discourse without texts that exhibit it empirically, and there are no texts without there being a discourse to develop. There is no primacy between the two phenomena, only necessary complementarity. The reciprocal presupposition between texts and discourse sets in motion the fundamental dialectic between an Enunciator, which produces texts in order to spread its own discourse, and an Enunciatee, who, starting with the texts that are proposed to them and the other possible texts circulating near them, reconstructs the discourse in their own way. It is a somewhat crooked dialectic, because between the discourse produced upstream and that reconstructed downstream there cannot be a total similarity, an absolute overlapping. Rather, it depends on the fact that between the discourse upstream and that downstream, there is the mediation of a series of texts that is not shared entirely between the two actants in the communication. Only some texts produced by the Enunciator will be received and understood as such by the Enunciatee, who will consider other texts pertinent, produced by other enunciators. The producer's point of view and that of the audience do not coincide.

This makes it extremely difficult to determine *a priori* if and what contributes to the overall construction of a discursive field, as well as *a priori* decisions regarding the physiognomy of a text. Textual jumbles are continually negotiated and renegotiated, sometimes finding convenient forms of agreement, sometimes activating more or less violent conflicts in a discursive flow without a beginning or an end. The question of the formal constitution of textuality and that of the discursive constitution of textual series become the same, so that the *text and inter-text*, different *a priori*, deriving from the same communicative processes of constitution and circulation, *should be understood* a posteriori *as the same thing*. For example, an advertising campaign upstream will be composed of a series of spots, press releases, billboards and so on, each of which is in itself a closed text that, placing itself in relation with others, takes on a precise signification. However, downstream, not all of these texts arrive at their destination, something that will inevitably modify the signification of each of these. Or, if a given promotional event is held at a certain outlet, two different texts (the shop's space and the event in itself) become a single one (the situated event) (Marrone 2007). Understanding the confines of the text means understanding its links with other texts and establishing bonds between texts means determining the physiognomy of each of them. In the daily practice of the production and reception of discourse, all this depends on situations of communication and the circumstances of reception. And all this depends (in the practise of whoever intends to analyse all this) on the semantic dependencies that they intend to expose in order to reach the stated hermeneutic objectives. The problem becomes one of the construction and maintenance of *discursive coherence* starting with very different kinds of texts. The texts that contribute to the manifestation of the discourse can use the same expressive languages (verbal, visual, audio, gestural, etc.), which are often mixed together in varied syncretisms (comics, audio-visual content, songs, bodies that move in physical spaces and so on). And, in this case, the forms of coherence will be *intra-semiotic*. But the texts used to express a self-identical discourse can use different expressive languages (as well as different media), demanding more complex and delicate forms of coherence that depend on *inter-semiotic* (or *inter-media*) translations. An example occurs when it is necessary to ensure the meaning of a spot corresponds to that of a shop, that of a product corresponds to that of a promotional event, and so on. Another example might involve a website juxtaposes very different textual substances and modules to form an entirely new hybrid that gives rise to new forms of promotional communication and ways of calling upon the receiver. One of the most effective ways of producing and maintaining this discursive coherence, starting with texts of a substantial and mediatic nature, is that of storing on the surface not only the same profound values, but also the communicative pact itself. By endowing the Enunciator and the Enunciatee present in various texts with the same modes of action and passion, as well as a shared form of reciprocal dependency, it is possible to keep constant control over communication and its strategies. Thus, a public institution or a private business, having to express themselves with different means and languages, and speaking about rather different things, work on an enunciative level in order to construct and maintain their own identity, placing themselves within different texts: someone who always says 'I' in the same way.

A very different form of coherence, which we will discuss in the next chapter, is that of an eminently aesthetic kind, that is visual, or derived from other modes of physical sensoriality (smell, taste, hearing, etc.), and that is bound to the level of expression with its secondary significant outcomes. In this case, the body takes charge of discursivity (as we will see), so that identity no longer passes through a cognitive (recognitions, comprehensions, etc.) or passionate (identifications, enthusiasms, etc.) dimension, but through the activation of somatic meaning.

Halfway between these two is the typically discursive coherence punctuated by the array of semantic categories (themes and figures) and syntactical ones (actors, spaces, times), as well many more configurations that their possible combinations can create. The theme of a discourse is the realisation of underlying narrative values that are further actualised by a series of *figures of the world* that render it possible. Figures that are structured among themselves in totalities that are sometimes *partitive* (a paratactical accumulation of images linked by the same semantic principle, i.e. from 'tree' come figures such as 'pine', 'oak', 'elm', 'olive', etc. in an open and never-ending list), sometimes *integral* (semantic configurations internally articulated by some form of syntax, i.e. from 'celebratory meal' come figures such as 'fine china', 'silver cutlery', 'elegant clothes', 'exquisite food', etc.). This semantic nexus between themes and figures is constructed thanks to connections and dependences between *actors* (realisation of actants), *spaces* and *times* (realisations of narrative programmes) are established within a discursive configuration. A 'celebratory meal', as well as having stereotypical figures, possesses typical characters, its own canonical temporality, standardised locations, and so on. The coherence of the discourse is constructed (horizontally) thanks to the presence within the same textual package of figures belonging to either one or other forms of figurative totality (as well as the recurrence within them of the same actors, spaces and times). And it is constructed (vertically) thanks to the progressive realisation and enrichment of an abstract story. First of a precise theme, and then in a specific form of 'figurativity' (which we will discuss in the next section), and a clear array of actors, spaces and times, which we will discuss later.

IV.6 Themes and figures

We usually assume that linguistic terms have a *literal* signified and a *figurative* one. The first would be its actual signified, the one that we find in the initial definition given by dictionaries, and refers to the referent external to the sign. The second would be an added signified, from the unique virtuosity of the singular speaker or writer, that does not refer to a real external thing but indicates something different that enters into common usage, giving rise to clichés or stereotypes. So, on the one hand, we have the 'rational' meaning of the sign, constructed firmly in the common language and, on the other, its 'poetic' meaning, the result of a creative imagination or constructed ad hoc in order to extend, on specific communicative occasions, language's potentiality. This vision of semantics that derives from ancient rhetorical tradition has been broadly overtaken by contemporary linguistic and semiotic reflection. It has been demonstrated that the meaning of terms is in no way stable but transforms over time. What is illustrated by dictionaries is, therefore, only a cross section of the synchronic situation of language that is forced to treat its diachronic evolution as an afterthought. Secondly, the linguistic signified is not a concept, something unitary of a cognitive-rational kind, but is composed of diverse entities that are logical, visual and, generally, sensorial. Thirdly, there is no signified linked to a single term, because linguistic units comprise vast phrasal and textual entities that produce signification.

For these reasons, it is worth thinking of the meaning of words in relation to a configurational signification bound to the complexities of discourses, to tangible communicative situations, to cultural and social contexts, to historical eras: in short, to texts. Furthermore, this signification holds within it elements of different kinds: some are interoceptive, *abstract*, intellectively constructed entities (such as the oppositions thing/event, process/system, and so on), some are exteroceptive, *figurative* entities reconstructed through the use of the senses (such as volume/surface, light/dark, sweet/bitter, etc.), some are proprioceptive, somatic entities that articulate the *timic* category (that opposes, as we know, euphoria to dysphoria). In other words, the elements that contribute to the construction of signification are not intellectual, devoid of any contact with the external reality, but often pick up their sensitive qualities from reality, i.e. from the properties that subjects perceive through their own senses. Oppositions such as right/left, sweet/bitter, shrill/smooth, that are already significant in our everyday world, are used by the various systems of signification to construct their own plane of content. Thus, an inter-relationship is created between the macro-semiotics of the language and the world that reinterprets the relationship of reference in terms of translation.

Projected onto the model of the generative path of meaning, this issue finds a position on the level of the semantic component of discursive structures, where the values are produced at the level of the semiotic square and the narrative articulations at an anthropomorphic level, through the mediation of the subject of enunciation, are thematised and depicted. On the basis of the generative path of meaning it is possible to propose a progressive realisation of the semantic elements of a text in an incremental way, from narrative structures to discursive ones. So, a narrative structure in which an actant Subject will go in search of an Object in which the value 'freedom' is inscribed, and in which modalities such as want-to and able-to will spring into action, is manifested on the semantic level of the discourse through a virtual series of possible themes. If the discourse is journalistic, it will give rise to themes that are political (federalism, presidentialism, autonomy of parties, etc.), social (equality between classes, races, religions, regions, etc.), economic (right to pensions, fight against inflation, pay rises, etc.) and so on. If the discourse is literary, the idea of freedom will be rendered in a more substantial way, giving rise to a range of themes that go from evasion to revolution, as well as liberation from a social or spiritual constraint. Thematisation is, in short, the semantic covering of narrative structures through the selection of a series of possible themes.

Every theme can be depicted in different ways. The search for freedom thematised as federalism can evoke figures such as green shirts or flags (in Italy), or the stars and stripes (in USA). Similarly, the theme of the social state can be 'figurativised' in the form of strikes, blocked streets, chaos on city streets caused by traffic jams (in a novel), or in the form of ordered queues of old people lining up at the post office to collect their pensions (in TV news). There are themes that are connected automatically with certain figures so as to create real *discursive stereotypes*. For instance, the arrest of a fugitive evokes police cars shooting along crowded streets, people getting into cars trying to hide their faces from the television cameras. The newspapers are full of these kinds of fixed combinations of themes and figures that have quickly become stereotypes, even on the basis of a widespread media imagination taken from cinema, television and advertising. Just think of how the news of the ISTAT[1] statistics on the lives of Italians are visualised: people at the market, piles of money counted out by invisible hands, people walking through city streets, shopping or otherwise. But, if we change genre, we can think of how war in third world countries is figurativised: images of people searching for water in remote wells (or extreme close-ups of children with flies in their eyes, or lying on the floor, moribund) are interspersed with other images of soldiers shooting at invisible targets (or someone emerging from their hiding place in search of another, or throwing hand grenades).

While the theme unites narrative semantics with its figurative expression, figures constitute one of the highest steps on the generative path of meaning: a level of signification where the tangibility of the world is presented in all its variety and complexity, hiding the underlying abstract structures. The sensitive qualities of the natural world, the empirical properties of things that are significant for us are gathered within the discourse, contributing to the constitution of its forms of content. In this way, the discourse enriches itself with signification and consequently produces the image of a reflection of reality. As they are already significant in the natural world, figures are never neutral. At the moment in which a certain arrangement of images is created in order to exhibit certain themes, we must reckon with what these images already signify in the myriad other texts in which they are used. One example: if in order to exhibit the theme of federalism I choose the figure of the stars and stripes, I am referring to federalism in the United States, but I am also, necessarily, calling into play everything that the United States might express, and everything that the stars on their own and the stripes on their own can, in turn, reveal. And, again, if in order to exhibit the theme of serenity I choose to evoke a half moon, I could suggest, willingly or otherwise, Islam, whose symbol is a half moon.

The relationship between themes and figures is two-sided. If, on the one hand, the theme is an abstract semantic reality that is figurativised, enriched with meaning through an array of figures, the figures are carriers of different themes. What we might consider accessory elements to a story on the level of narrative structure are in no way irrelevant when considered on a discursive level. For example, the flow of fairy-tale 'motifs' can take place through a thematic recurrence (with a modification on a figurative plane), or with a figurative

---

[1] ISTAT is the Italian National Institute for Statistics.

recurrence (with a modification on the thematic plane), rather than with an automatic re-proposition of the nexus theme-figures. Furthermore, all the theory of so-called rhetorical figures finds here is its correct theoretical placement: the 'decorations' of the discourse should not be considered to be additional frills of abstract concepts, but as constitutive elements of the discourse, necessary to its semiotic organisation. The rhetorical level of discourse, then, is not lexical or phrasal but distributed throughout the text. So, a narrative genre such as the parable sets in motion discursive mechanisms similar to those used in lexical metaphors. From this point of view, parables are narrative metaphors, tales that metaphorise other tales, stories in which the thematic-argumentative level and the figurative-poetic level slot together. Figurativity generally tends to detach itself from its thematic base and render itself relatively autonomous, producing *figurative reasonings* that overlap the text with further significations.

It should be made clear that the level of figurativity must not be confused with the plane of expression in visual languages (which we will discuss in the next chapter), but features in the plane of content. The figures of the world that render tangible or carry possible themes are one thing, the images that make those figures tangible by appealing to a precise substance of expression, such as visibility, are quite another. In fact, figurativity can be rendered through words (as with verbal description) and (why not?) through music (take the commentative function of certain soundtracks in action films or spots). And while it is clear that figurativity finds a more appropriate manifestation in visual substance, there are cases in which even verbal language can render the figurative plane of the discourse admirably.

From this comes the idea that, within discursive semantics, various sub-levels of figurativity exist. One thing is the basic figurative elements like high and low, right and left, while figures such as air, water, earth and fire are quite another, and the figurative details of characters in a story are another still. Even if all three of these cases can be considered expressive components of the natural world called into the discourse with the aim of figuratively fulfilling its abstract structures, there is no doubt that they are profoundly different. And they too must be placed on different levels of profundity. The category high/low, though figurative, is sensitively more abstract than so-called Aristotelian elements, which are more abstract than their textual determinations (river, sky) or the black colour of Madame Bovary's hair. It is necessary within the figurative level of discursive semantics to foresee some sub-levels, ordered hierarchically according to increasing *figurative density*: in the first level, known as *figural*, a few figurative formants begin to cover the thematization; in the second, the *figurative* level, the world's first complete figures appear; in the third level, known as *iconic*, figures are gradually enriched with increasingly minute details. We will return to this when we discuss images.

The distinction between the thematic and the figurative levels, and, within the latter, the separation between the figural, figurative and iconic sub-levels, allow us to draft a typology of discourses based on formal semantic criteria rather than heterogeneous principles such as, for example, the traditional partitions found in disciplines or many systems of literary genres. Discourses are created in which the thematic component is privileged (as in much philosophical debate) and there are discourses in which the figurative component plays a primary role (as with much literature). Within the latter, the kinds of discourse in which a tenuous figurativity prevails distinguish themselves from others in which iconicity is developed through the use of various kinds of stratagems. In this last case (which can, as we have already said, be either verbal or visual), the text, enriched by identifiable figures of the world (known as *icons*) are charged with *effects of the real*. This means that the receiver, through a specific reading framework, is led to recognise a number of figures from the natural world that they had previously identified in those icons constructed within the text. As such, the receiver tends to consider the message contained in that text as something true. This is the reason why, for example, photography is generally iconic, when compared to drawing, because is able to generate truthful effects of meaning. In the same way, a short story, perhaps one by Balzac, in which characters and situations are described in minute details, is considered much more truthful than another short story (perhaps a fairy tale), in which the characters do not have proper names and whose actions are not rigorously contained within a spatial-temporal limit. Realism, be it literary or otherwise, is not created from a preliminary poetic project whose works are simple actualisations, but is the result of the figurative recognition by the receiver of specific textual configurations.

Recommended bibliography

On discourse and discursivity: Barros (1988); Brown & Yule 1983; Coquet 1985, 1997; Duranti & Goodwin (eds.) 1992; Fontanille 2006a; Gumpertz 1982; Halliday 1978; Levinson 1983; Lozano, Peña-Marin & Abril 1982; Manetti & Violi 1979; Parret 2019; Rabatel 2017; Violi (2019.

On enunciation in philosophy, speech acts and pragmatics: Austin (1962); Ferrara (ed.) 1980; Goffman 1981; Wittgenstein 1953.

On enunciation in linguistics: Benveniste 1971, 2015.

On enunciation in semiotics: Barros 2919; Colas-Blaise, Perrin & Tore (eds.) 2016; Coquet 2007; Fiorin 1996; Paolucci 2020; Lancioni & Lorusso (eds.). 2020.

On efficiency and efficacity: Freedberg 1989; Gel 1998; Gombrich 1999; Leone (ed.) 2014; Pezzini (ed.) 2001; Sherzer 1983; Volli 1997.

On the cognitive dimension of discourse: Fontanille 1987, 1989; Greimas & Landowski (eds.) 1979; Pozzato 2001.

On intertextuality and intermediality: Badir & Roelhens (eds.) 2007; Bertrand & Estay Stange 2017; Bolter & Grusin 1999; Hebert & Guillemette (ed.) 2007; Peverini & Spalletta 2009.

On themes and figurativity: Basso Fossali 2017, 2019; Bertrand 2000; Floch 1985; Geninasca 1997; Gombrich 1951 (on figurality); Hamon 1980 (on literary description); Moreno 2019; Pezzini 2019.

V Image, sensoriality, body

V.1 Forms and substances of expression

Having examined all the levels of the generative path of meaning with regards to narrativity and discursivity, we will now focus on the mechanisms of textualization and the ways in which the plane of content structures itself as a text by attaching to a plane of expression. On the one hand, the plane of expression has always been present in our considerations: the very act of drawing the elementary structures of signification within a square, for example, gives that theoretical model a precise expressive quality. Not to mention other levels of meaning, which in order to be deliberated must be said and expressed somehow. On the other hand, it is also true that taking a particular form and substance of expression means that the text can further increase its own meaning, enriching and transforming it. For example, a verbal tale is one thing, a visual one is another, an audio-visual one another again, and the same goes for comics, music and so on. Describing a character with words requires a certain effort, whereas with an image this can be done much more efficiently. A film cannot avoid the problem of how a certain person or environment looks, but a novel can. Similarly, depicting someone or something in a painting or photograph does not involve the exact same operation, much like how a passion in music is not expressed in the same way as it is in a drawing. In short, though the text (and whoever is analysing it) is able to place the plane of expression to one side when it comes to its non-immediate or obvious levels, those of narrative and discourse, there comes a moment in which that plane becomes pertinent. On this level (and only on this level), language, music, painting, photography, comics, architecture, sculpture, etc. all become specific codes with their own forms and substance, which, by interweaving with one another, help to enrich the text's plane of content, multiplying their own meaning.

As for the forms of expression, specific disciplines exist to deal with these. Linguistics has been studying the constitutive mechanisms of verbal texts for centuries. Similarly, for the other kinds of text there are disciplines that study the semiotics of music, cinema, photography, painting, architecture, design, urban planning, advertising, information, and so on. Likewise, when it comes to substance there have always been techniques or artforms such as poetry or poetics that play on language's capacity for sound or musicality, in the same way that, in the field of the visual arts, there are those that specialise in studying light, colour, or different materials. The role of semiotics in this context is one of integrating all of these formal and substantial elements, demonstrating how their signification is only realised in precise configurations of meaning that are texts. Anything – sensitive qualities, perceptive events, the consistency of substances, the nature of materials, portions of physical space, etc. – can be a potential vehicle of meaning, provided that it enters into a pertinent relationship with other, equally significant elements within specific textual manifestations.

In this chapter, we will consider the text from the perspective of its expression, noting how even those substances we use to express ourselves (visuality, sensoriality, corporeality, spaces, etc.) are often all carriers of further, specific meanings. New aspects of textuality will emerge, such as those related to the image (too long considered to be somewhat, if not entirely insignificant) and also to other forms of sensoriality (sound, smell, taste, etc.), and as such, the body as an essential site for the production of meaning, plus space as its essential alter ego.

V.2 Figurative and plastic

Having overcome the long discussions on so-called iconism, i.e. the natural or conventional nexus between images and reality, semiotics insists on drawing attention to the *doubly signifying nature of the image*. On the one hand, the image signifies thanks to perceptive cultural grids (as the image represents something within itself) and it is the manifestation of a series of figures of the world that, by referring to a visual expressive substance, are placed on their own plane of content. A drawing, a photograph, a painting, a cartoon, a poster or a logo often 'represent' objects or people, landscapes, things, artefacts. On the other hand, an image can be the carrier of further meanings related to its specifically visual aspects, known as *plastic* (using the terminology from the French language that is now commonplace). For example, by articulating with one another, colours can give rise to certain signified elements; graphic or pictorial forms can likewise become the carriers of a specific meaning. Not to mention the disposition of visual objects distributed throughout the surface of the graphic format, or the *textures* of an image that involve touch and lead to further signification. And that goes for both artistic depictions (valorised as such in different cultures) and depictions whose objectives are instead illustrative, journalistic, documentary, scientific, for advertising purposes and so on.

In other words, the visual substance in an image signifies twice because it is formulated twice.
(i) Initially (in a figurative instance), the forms, colours, materials, and so on that are contained within an image signify because they tend to mimetically reproduce something that we already know from our own experience, to

the point that, usually, when we look at an image, we neglect its signifying aspect in order to concentrate directly on its signified. When looking at the photograph of a lion, for example, we will say simply that we can see a lion and not its photographic reproduction obtained using a particular technique, particular lighting conditions, a particular lens, focus, and so on.

(ii) Unless (and this is the second instance: the plastic), for various reasons, the visual technique utilised does not jump out at us, meaning other possible messages are communicated. The sepia-tinged photo of a person, as well as depicting them for us, tells us (for example) about the time that has passed since the photo was taken, and therefore about the past that no longer is, and so on. Similarly, a medieval painting with a sacred subject, as well as representing a particular biblical scene, tells us about the hierarchy of the people who appear in it according to their size and placement on the canvas: the sacred characters are large, often appearing at the centre of the image, whilst those who are secular are smaller and appear lower down. Take also those advertising spots that play on the passage from a black and white colour scheme to one featuring colours, associating negative values to the former and positive values to the latter. In these cases, the transformation from black and white to colour, which usually occurs with the appearance of the product being advertised, takes on the representation of the narrative transformation and the conjunction of the Subject with the Object of value. Obviously, this does not mean that black and white is, in itself, a dysphoric symbol in all advertising spots and that colour is, as such, a symbol of euphoric values. That would depend on how the audio-visual substance acting as an expressive format is used. It is worth drawing a parallel here: in the case of language, the sound used on the signifying plane in order to transmit particular signifiers is arbitrary, apart from those cases (such as poetry) in which this takes on further meaning (in a banal rhyme such as *care/ share/ fair/scare*, putting words together because of their sounds furtively suggests that we also associate them because of their meaning). In the same way, the signifiers (shapes, colours, positions in space, etc.) charged with representing particular figures can become carriers of new signifieds described as *plastic*, if they are *articulated in another way within the same text*.

V.2.1 Time in images

One of the most common and obvious cases of this double language of images is that regarding the visual representation of narrative time. Images are static, and so they are not supposed to represent multiple moments from the same tale. Either many cartoons are used, as with comics, following on from one another from left to right and from top to bottom (according to Western conventions), or a single image, which focuses on a single moment from the whole tale, perhaps the most significant, in the hope the spectator is able to complete it with their own prior knowledge. And yet, much of art history (and not only) suggests the opposite. Well before comic strips, artists had thought up different ways of overcoming the problem by including several moments from the same tale in a single painting: placing, for example, an earlier moment on the background and the final moment at the front or, alternatively, the former on the left and the latter on the right, or even the former at the centre and all the others around the edges. The space of the painting thus signifies twice: once for what it represents (roads, buildings, landscapes, environments, etc.) and a second time through the different positions (right/left, centre/edges, etc.) occupied by the various figures within the canvas. Just think of the sacred representations of the crucifixion, in which very often in the foreground (in the centre of the painting) there is Jesus on the cross with the two thieves, while in the background (so smaller and higher up on the canvas) is the same Jesus carrying the cross up to Calvary. From this come the analogical relationships "high/small: first = central/large = after". Not to mention the predellas of the altarpieces, those long, painted strips that, beneath the larger paintings, tell the life of a saint, representing the most significant moments of their sacred existence from left to right.

And again, to give a specific example, take the very famous work *The Tribute Money* by Masaccio (1442) (see Fig. 11), in which the central scene depicts two earlier moments (the tax collector asking for Peter's tribute [1], Jesus and Peter pointing to where the money can be found [2]), the scene on the left is the following one (Peter finds a coin inside a freshly caught fish) [3]) and the scene on the right is the final one (Peter gives money to the tax collector [4]).

Let us look at an example from advertising (see Fig. 12). Here, the simultaneous presence of these two languages – figurative and plastic – within the image is more obvious. In the French advert for the anti-anxiety medication Sédatonyl, the communicative content emerges immediately: thanks to the product, the subject goes from being in a state of anxious dysphoria to one of euphoric calm. How is this message produced? By using an enormous number of expressive categories that are both figurative and plastic. Figuratively, we see the yoga position of the person at the bottom, a well-known icon of relaxation, and a position that is, conversely, not adopted by the other two figures in the middle and at the top, even though we can tell that the character is moving towards that position (Floch 2001).

So, the calming process is determined figuratively by the three positions of the body: the arms are up, then they lower, before finally assuming the 'correct' position. The same thing happens with the legs and something similar occurs

Fig. 11: Masaccio, *The Tribute Money* (1420), Florence, Santa Maria del Carmine. Brancacci Chapel (with the different moments of the story)

with the facial expression. On a figurative level this narrative process is communicated also by the way in which the woman's silhouette, or rather the black line that depicts her outline, is drawn, in which the discontinuous line (due to convention) gives the impression of fear, agitation, anxiety, whilst the continuous one suggests an idea of calm and interior peace. On a plastic level, this calming process is overdetermined first and foremost by a series of exclusively visual categories: the placement of figures at the top, middle and bottom communicates the time of passionate transformation (hence the analogy 'high': 'low' = "before": "after"). Then, we have the dark and the light, which directly convey the timic transformation ('dark': 'light' = "dysphoria": "euphoria"). But the asymmetric and symmetrical figures are also plastic elements that, in their own way, reiterate that same passage from dysphoria to euphoria. Here, then, we have an excessive proliferation of figurative and plastic traits used to signify a concept that is actually very simple; and it is clear that these traits are not the same. What interests us here, for obvious reasons, is highlighting the semiotic difference between *figurative formants* (cultural symbols, the outline of figures) and *plastic formants* (colours, the positions of figures on a page, lines, shapes). Take the example of the crucifixion, in which the plastic formants (high, central) do not have any particular signified on a figurative level.

These examples allow us to make two observations. Firstly, that *the plastic plane of the image is not the plane of expression of its figurativity*, but a *second language* with its own expression and content. The plastic is therefore a language in every sense, one which overlaps with (often hiding beneath) the figurative. If this were not the case, we would not be able to guarantee ourselves the possibility of

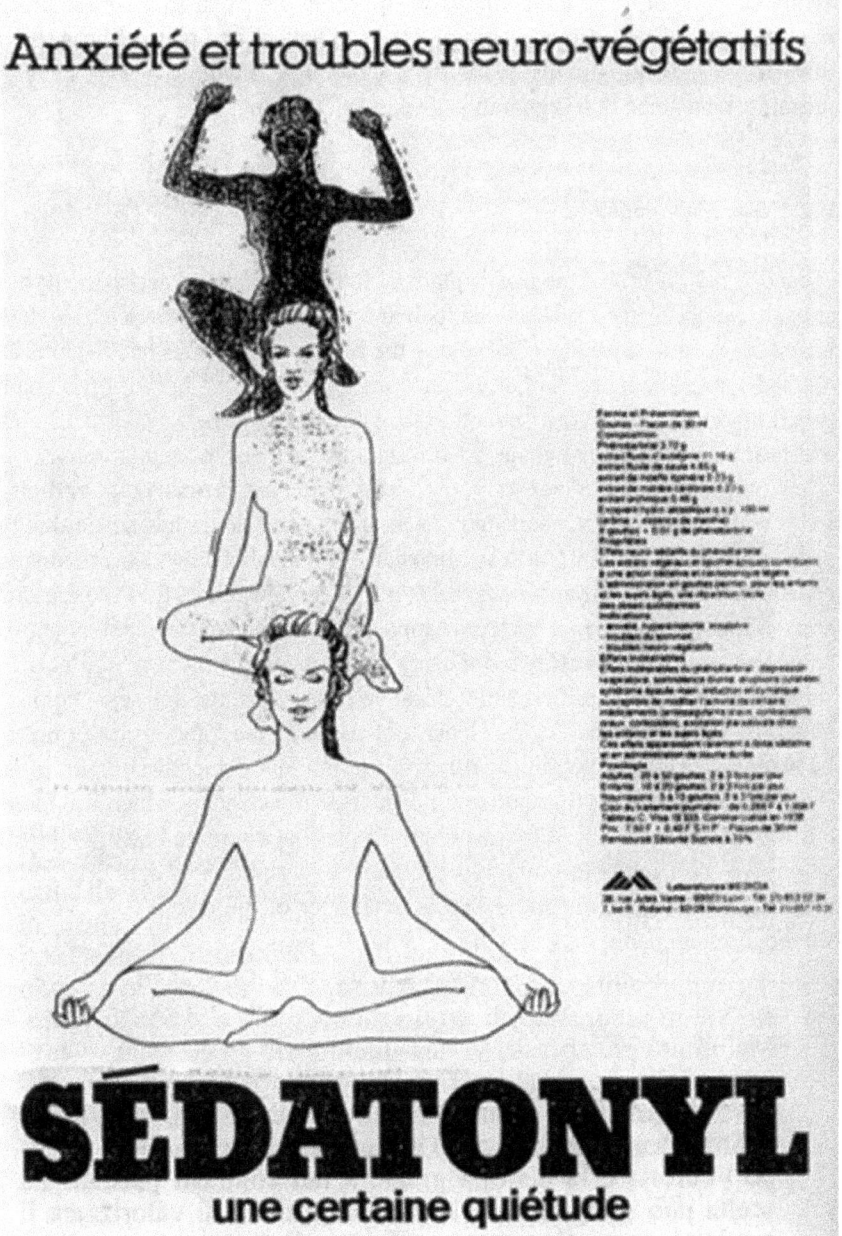

Fig. 12: Advertising of Sedatonyl, France (1990).

giving meaning or, most importantly, of explaining the articulation, of the images in what is known as abstract art (from Picasso to Pollock and so on). Those images that do not have a figurative plane do not represent anything despite having a meaning: a meaning that is guaranteed by the plastic language.

V.2.2 Semi-symbolism

Secondly, it is clear how the plastic plane of the image does not act on the basis of single symbolic elements (as can happen at a figurative level, where – as with the advertising example we have looked at – the woman's yoga position is a culturally defined symbol of calm). Rather, it functions using what we have (provisionally) called analogies: through a semiotic procedure called *semi-symbolism*. The elements at play in plastic language are not actually symbols, but semi-symbols.

What does this involve? Take the traffic light: red means 'stop' and alternates with green, which means 'go'. Each of the two colours has a meaning but only when the other signifies the opposite. Semi-symbolism exists when small code functions in just that specific case or in just a few others that use the same convention. In this way, at least two opposing elements on the plane of expression correlate with at least two opposing elements on the plane of content. The sepia-tinged photo signifies 'past' because those that are not sepia-coloured generally signify 'present' (or at least, not-past). In the religious paintings we have looked at, the large-scale central figures are hierarchically superior to any found on the edges or that are smaller in their dimensions, which are hierarchically inferior. Similarly, in advertising spots that alternate black/white with colour we can use the formula 'b/w': 'colour' = "dysphoria": "euphoria". What produces meaning is not the symbolic character of the colour, but the semi-symbolic relationship between the chromatism present in the same text (or in different texts that are in some way connected) and the meanings that each of these carries.

From the perspective of semiotic analysis, reconstructing semi-symbolisms is required to circumvent blackmail by culturally determined symbolisms (for example: of colours, shapes, positions, etc.) that can convey semantic inversions (is black a colour of mourning or elegance? Is height a symbol of authority or divinity?). However, it is also required in order to escape the vagueness of connotations, uniquely dependent on whoever (subjectively) notices them – if and when they do so.

V.3 Figurative reasoning

Now we have clarified the difference between the two languages of an image, we will attempt to examine them one at a time, setting to one side our knowledge of the communicative games that can occur thanks to their simultaneous presence in the same visual text or in the same body of images.

V.3.1 Representation and perception

What exactly is meant by the figurative plane? As you will remember, we have already discussed it when, with regards to discursive coherence in different textual manifestations, we recalled the nexus between thematic configurations and those that are figurative [cfr. IV.6]. While the thematic level of discursive semantics has a certain degree of abstraction and involves cognition and conceptual interpretation, the figurative plane is its concrete aspect and the way in which external reality is perceived. Figurativity is that level of meaning responsible for human and social perception of the world, primarily in a visual way (though, as we will see, it also involves the entire sensory and somatic apparatus). So, if 'escape' is a typical narrative theme, then there must be a figurative apparatus that makes it tangible, perceptible, 'realistic' somehow: an interplay between sawn-off bars, knotted bedsheets snaking down a wall, muted sirens, floodlights that illuminate the darkest corners of the jail, escapes into the night, etc. The way in which themes are depicted is in no way stable but varies from culture to culture, age to age, society to society, text to text. Therefore, a similar theme could lead to very different figurative representations. In this way, the figurative apparatus that undertakes to represent abstract themes has nothing natural about it, nothing that can be generalised, nothing universal or necessary. Such an apparatus depends on a perceptive process resulting from a 'reading grid' projected by every culture onto the world of their experience, in order to make it both perceptible and significant: significant because it is perceptible, perceptible because it is significant.

From this point of view, perception is not bound so much to our sensory, physiological apparatus as to the way in which every subject that perceives (and before them, their body) sits within a certain social, cultural and historical environment. Vision in particular is not a natural operation but depends, much like language, on a series of social codes that transcend it, and that, as with language are both arbitrary and shared. The constitution of a certain portion of the world as a figure (let's say 'house') derives from what we have previously understood by house: a dark cave, a straw hut, a paralleliped with a pitched roof and a protruding chimney, a city apartment, a skyscraper? *Seeing something means deciding in advance what to look at.* A figure, a group of figures, an event in which many figures follow on from one another at varying speeds, etc. are the result of this preliminary encounter between some visual form of the world and some sense of the world itself, between a system organised according to perceptive traits and another organised according to semantic traits. If, in order to represent a zebra, we draw something akin to a horse with stripes, in a population where only zebras exist, all that would be required would be the profile of a horse. *Images represent not things but our ideas of things.* This is the reason why, in an image such as in Fig. 13, we can see both a rabbit and a duck, but never both at the same time. Our eye focuses on the rabbit if we decide, as it were, to see the rabbit, if we look for it among the patches and lines on the page, making an effort to find it. This is also what we do with the duck, but in a different moment and mental situation that does not overlap with the first.

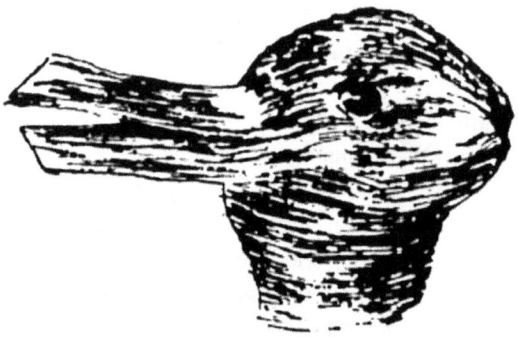

Fig. 13: Joseph Jastrow, *Optical illusion: rabbit or duck?* (1899).

So, when we say that a certain image is 'realistic' and another is not, we are implicitly bringing what we see in the two images together with the idea that we already have of whatever is being represented (the world that is 'real' to us) on the basis of a series of codes of both representation and vision. As far as we are concerned, the images used in this procedure are realistic from the moment the central perspective became the habitual way to represent the space, while those that are not used are perceived by us as strange, unrealistic, badly constructed and, as such, 'false'. If figurativity implies a certain degree of realism (leading us to say that images imitate 'reality'), such a mimetic effect derives from a kind of social agreement on what reality is, or rather, if we think about it, on what is significant about it, and on the conventional ways of rendering this significance. Dürer depicted rhinoceroses with scales not because in his time real rhinoceroses were not known and so he did not know that these animals do not have scales at all; he did it simply because he was using a centuries-old tradition in which rhinoceroses were depicted according to that figurative model. And, in order for his engravings to be considered realistic, he needed to ensure they adhered to a cultural model rather than a real animal.

V.3.2 Degrees of figurativity

Not all images are realistic in the same way. The mimetic effect is different each time, both on the basis of cultural variations and how the image is created, which can be varyingly dense or rarefied depending on the quantity (and quality) of figurative traits it uses. Generally speaking, the richer a visual text is in visual traits and details, the more realistic that text is considered. And, conversely, the fewer visual traits a text has, the less realistic it is considered. A circle on its own is a circle and nothing more. But if a series of traits are progressively drawn on and around it, the circle becomes a sun or a ball or a face or any other possible figures. Conversely, if we progressively remove those traits from the image of a face, say, it will at best remain a circular figure and therefore no longer a figurative image but another abstract one. *Emoticons* and their graphic transposition using certain letters and punctuation marks follow this same principle.

This brings to mind the famous example of a ball that, through the addition of small graphic traits, takes on very different semblances, becoming first a shopping bag, then a coin purse, then a cat (see Fig. 14).

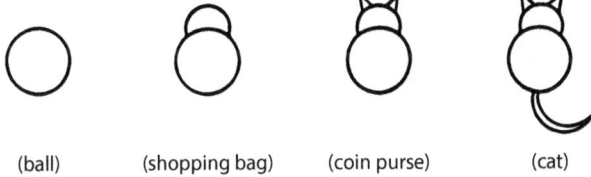

(ball)   (shopping bag)   (coin purse)   (cat)

Fig. 14: Different degrees of figurativity: how to draw a cat?.

If we wanted to generalise, we could say that the figurative level of the discourse allows for a kind of sliding scale that goes from maximum figuration to maximum abstraction, passing through a series of intermediary degrees. In order to establish some parameters, semiotic theory chooses to focus on three distinct sub-levels of figurativity:

- The first is the *figural*, in which few figurative formants begin to be drawn without yet taking on easily recognisable configurations;
- The second sub-level is the *figurative*, where certain figures from the world of our experience become recognisable;
- The third is called *iconic*, and is where these figures are enriched with increasingly minute details through progressive additions until it becomes a standard, canonical interpretation of the figure.

For example: a paralleliped is one thing (figural), a box is another (figurative), a strongbox is another again (iconic). Just as in the case of the ball that progressively becomes a cat, the difference between the images does not lie in the things they represent, in their ostensibly empirical reference points, but in their various *figurative densities* that show the spectator different things.

From all of this comes a series of relevant considerations. First and foremost, it is clear that figurativity is not exclusive to images but can be found (in different ways, obviously) in any other semiotic system, including verbal language. Indeed, verbal meanings have always a visual component, known as *figurative semis*, that helps to produce effects of meaning in words as soon as these words are inserted into a given linguistic context. The result is that the sub-levels of the figurative plane (figural, figurative and iconic) are also present in verbal language. I can say "trousers" to define this word or say, as the dictionary does, "an outer garment covering the body from the waist to the ankles, with a separate part for each leg". Or I could talk about a particular kind of trousers, describing their style, their provenance, the steps in their production, break them down into their smallest details, and so on. In the first case, I am using figurality; in the second, I am using figurativity; and, in the third, iconicity. And it is fairly clear that, once again, the more we charge a verbal description with figurative semis, the more 'visible' we make the object through the use of the procedure known to classical rhetoric as *hypokyphosis*, the more effects of realism are produced in the enunciatee ("what you are saying to me corresponds with what I usually see").

These sub-levels, however, are not only present in images and words in an objective way. They often depend on the kind of culturally-situated gaze projected onto the same image. In one particular painting, a spectator might see the figure of a woman whilst, in another, the iconographic elements (the blue veil, the halo) reveal her to be the Virgin Mary, and another again, by activating the intrinsic signified, will mean she is recognisable as (for example) the Madonna del Cardellino. It is not that the first gaze is, strictly speaking, wrong. Rather, it is simply naïve, unconnected to the image's codes of recognition that, in this example, make it a sacred image. This brings to mind a series of cases in which a kind of elementary uncertainty emerges regarding what the image represents, allowing it to be read in two ways. It also gives us a way of using images for humour, exemplified in a comic strip such as this in Fig. 15:

— It was a mistake to come on holiday to the mountains. It reminds him of the sales charts!

Fig. 15: Comic strip from *La Settimana enigmistica* (1994).

It's possible to read the line in the background in two ways: first, as a mountain range; second, as a sales chart. And it is difficult to establish what comes first on a perceptive level: is it the mountains that, in the eyes of the poor soul in the cartoon, are reinterpreted as a sales chart, or the line that is enriched in the mind of the two women, and the spectator, as mountains? It is difficult, and useless, to give an exact answer to this question. It depends on an individual perceptive choice. What interests us is the fact that the possibility of such a perceptive *switch* is due to the two sub-levels at play: one that is truly figurative (in which we read the mountains), another that is figural (the line) that, underlying both figures, allows for the perceptive passage from one to the other and vice versa.

From here, it is possible to uncouple figurativity from its profoundly semantic basis, making it autonomous and no longer using it simply to depict tales or underlying themes, but to install *figurative reasoning*. These forms of reasoning do not follow the standard logical procedures (deduction, induction, abduction), but employ various kinds of perceptive games, through more or less explicit references within a number of images, or even within the same image, that are sometimes figural, sometimes figurative and sometimes iconic.

Often, two or more figures are placed together because they have a similar silhouette (one figural trait that, like poetry, 'rhymes' with the other). In doing so it is suggested that they also share a meaning, at least from a certain point of view. And so, in an advert (see Fig. 16) the woman and the perfume bottle have the same shape. The bottle is even dressed in the same bodice as the woman.

Fig. 16: Advertising of a Jean-Paul Gautier parfum (France, 2004).

Somehow, the bottle 'is' the woman; or, likewise, the woman 'is' the perfume. In one way or another both are Objects of value: the woman for the sailor to whom she shows herself, the bottle for the woman who uses it. Or is it the other way around? Examples such as this abound in advertising (Marrone 2007).

V.4 Plastic language

From the relations between the sub-levels of figurative language emerge *forms of reasoning* that are both verbally unsayable and extremely clear to the observer: hidden to those who do not know or want to see them, obvious to those who, curious or cunning, want to go beyond the veil of appearances. But most importantly, difficult to attack by those who do not have the descriptive metalanguage with which to describe them, and therefore an alibi for those who want to imply content without openly declaring it. So, despite those who maintain the communicative immediacy of images, semiotics insists that images insinuate rather than assert, suggest rather than declare, hide rather than reveal, reason rather than represent. By calling upon a careful and active gaze, they essentially end up invoking cognitive activity. All of this becomes more clear if we move from an examination of the figurative component to an examination of the plastic one. Plastic language, as we have said, is a second language that overlaps the figurative, giving rise to specific formants charged with transmitting original meanings.

V.4.1 Plastic categories

The forms of this second language, which can be translated using the metalanguage of textual analysis, are multiple and complex. Among the categories pertinent to the construction of plastic language it is worth remembering:
– Those that involve lines and shapes (*eidetic*);
– Those that involve colour (*chromatic*);
– Those concerning the position of the figures within the contextual space being used (*topological*).

To these three categories we must add:
– *Light*, which combines matters of colour with matters of shapes;
– *Texture*, which partially falls into the realm of enunciation and partly into that of tactility.

For the topological categories (centre/edges, right/left, high/low), it is necessary to clarify that these should not be confused with the spatiality represented on a figurative level. This involves the issue of the various planes of the image, and therefore the nexus figure/background. The topological categories, however, come into play when the way in which the figures are positioned, for example, in an advert, on a canvas or a tapestry, becomes pertinent. So, in the Sédatonyl advert we have already discussed, the three characters are placed along the page's vertical axis, which is used to signify the temporal advance towards euphoric calm. A semi-symbolism is activated in which, as we have said, high is to low as before is to after.

As regards the chromatic categories, we have already given examples of the passage from b/w to colour in many advertising spots which activate a semi-symbolism in order to signify the apparition of the product or the brand's positive values. In the Sédatonyl advert, the timic transformation from dysphoria to euphoria is communicated not just through the topological category of high/low, but also the chromatic category of dark/light.

As for the eidetic categories (straight/curved, unbroken lines/broken lines, convex/concave, etc.), we can see in the same advert how the passage from dysphoria to euphoria is overdetermined by the opposition between the broken graphic line and the unbroken one.

V.4.2 Classique and baroque

The best way to understand plastic language is by watching it at work in the construction of any text involving fashion, visual or otherwise. Fashion is, in many ways, a visual phenomenon, dependent on the plastic organisation of shapes, materials and colours. Beyond its social and cultural value, or that tied to marketing and economics, what constructs the essence of fashion is the plastic dimension.

The case of Coco Chanel, a designer who worked on two levels, is particularly famous in this regard. On the figurative plane were the famous identificatory elements for this clothing brand – the trouser suit, the low-heeled shoe with a black toe, the quilted handbag, the bow, the golden buttons with the brand's logo featuring the double Cs – all tend to produce the idea of a modern, free woman who can be put to the test and win in the world of work, without losing her femininity or elegance. Yet, from a plastic point of view, what emerges are no longer the clothing elements used by Chanel, their combinations or transformations, but the silhouette all this manages to create, the *total look* that has remained constant over time, making it a fashion classic.

When we talk about 'classic', this term should be understood in a very precise technical way that sees it in opposition (according to theorists of pictorial images) to the 'baroque' (Wöfflin 1915). In this sense, Chanel can be defined in this way because it uses the visual characteristics of classic style that, when combined with those of a baroque style, have been classified in the following way:
- the *linear* (as opposed to the pictorial);
- the *distinction of planes* (with regards to the valorisation of depth);
- the *closed form* (in opposition with the open form);
- the *multiplicity of elements* (and not the diffusion of unique masses);
- the *immediate clarity of the forms* (as opposed to a light that shows a reality in thrall to its whim).

No matter the single items used or the way in which they are combined with others, the general form of Chanel's *total look* produces a sense of closure. On one side of the body, the shoe's black toe produces a rupture with the earth; on the other, the choice of short hair, a very structured boater or a bow tends to clearly mark a perceptive boundary between the top part of the figure and its background. Though the form is closed, this does not mean that a mass effect is created. The dominance of linearity (think of the suit's outline, the precision of the folds, the very defined collars and belts, the printed fabrics, the rigidity of the dress's drop) tends to produce an effect of multiplicity, making each of the single elements used to create the *total look* easily recognisable. If masses are present in these (for example in the accessories) they are localised and easily identified: baroque elements that, isolated from any possible baroquisme, end up highlighting the fundamental visual clarity. Finally, Chanel's classic style is also assured when it comes to light: the dominant colours (beige, red, black) and the fabrics (jersey, tweed, crêpe) allow for a complete control of the light, which is captured in order to avoid it catching the various parts of the *look* in the wrong way, accidentally transforming it (Floch 2000).

Classic and baroque, understood as styles that depend on plastic language, thus become, with their sub-categories, a strong model for describing and interpreting not just pictorial and photographic images but any aspect of visuality, be it fashion or any urban reality. In the latter, for example, the articulation of spaces (as we will see later on) corresponds first and foremost to the category that places *continuous* and *discontinuous* in opposition, which enters into a relationship with the ways in which certain spaces are inhabited with the forms of sociality that arise from this. This leads to the presence of spaces that are open and closed, single and multiple, superficial and profound, light and dark, and so on.

V.5 Sensory experience, corporeality, space

The visual dimension is not the only one that is significant on the plane of expression of a text. Hearing, smell, taste, touch and the entire poly-sensory experience rooted in the body are called upon to produce further meanings. Isolating the visual component from other sensory and somatic experiences, as we have done up until now for explanatory reasons, is inexact. Often, the gaze of the observer inscribed in the image is brought closer to the figure to the point of modifying not only the perceived object (which loses its form in order for its substances to emerge), but also the perceiving subject. They are invited to use touch, to touch what they are seeing, to express its material qualities. The pictorial genre of still life (see Fig. 17) often uses this *tactile* (or haptic) *gaze,* which is brought closer to the objects it depicts thanks to a play of light that highlights different materials and various consistencies (glass, metal, dough, fabric, fruit). Advertising tends to use the same strategy (see Fig. 18), managing, for instance, to communicate the vibration of a mobile phone by closing in on and blurring its outline.

Fig. 17: Pieter Claesz, *Still Life* (1628).

Fig. 18: Advertising of a telephone portable by Ericsson (Italy, 2002).

Looking is *a process that involves the totality of the perceiving body*, the entire sensory apparatus and the specific situation the body finds itself living. Every act of seeing presupposes an act of looking, a gaze directed by some intention, activated for some reason, directed towards an objective that is both corporeal and social, pre-individual and collective. The visual experience involves the totality of the sense. From this comes the need to work on the *aesthetic* dimension of meaning, where this term must now be understood in the etymological sense of *aesthesis* (= sensoriality), involving the body in its entirety and in all its complexity. The senses and meaning are phenomena that are closely correlated, if not actually the same.

The body is the place where, and through which, signification is constituted and reconstructed. Considering the body means thinking of it as an element and a process that are already social and that have a cultural destiny marked by precise interests, programmes and values, contributing to their formation or destruction. If human subjectivity has a corporeal basis, this does not mean that it is founded on an abstract naturality as a condition for the possibility of individual and collective diversifications. Human subjectivity is constantly constituting and reconstituting itself through pre-subjective experiences and inter-subjective instances. What is more, corporeality is not simply something that involves the senses and perception. The field of spatiality, that of so-called topological languages, closely relate signification and the body. We will come to this later on. Spatial structures, rooted in somatic ones, are highly significant for humans. On the one hand, space signifies for the subject (individual and collective) that sees and lives in a particular environment, reinterpreting the physical properties of that environment as signifying articulations on the basis of its own action plans and values. On the other hand, we have the space itself, with its signifying articulations, that contributes to the constitution of the subjective identity, allowing certain forms of action while impeding others, and for this reason presenting itself as an axiology. As such, semantic oppositions like high/low, right/left, in front/behind, internal/external are not physical distinctions that derive from the fact that the human body has a particular form and postures. In different ways, depending on the cultures or single texts, these oppositions become actual value systems that dictate possible plans of action and passion to the subject. Body and space are closely linked: just think of the experience of being contained, which is constitutively paradoxical because it is specific to a body that inhabits a space and to that same body equipped with its own specific spatiality that is, therefore, a container at the same time.

Thus, current semiotic research considers sensory and corporal experience as a kind of co-participant in the construction and transformation of signification. There is no semiosis if not with a body that lives, that experiences the world, that enters into a contractual or conflictual relationship with it, that constitutes it and is modified by it. The body is involved with the materials of expression of various languages – auditory, visual, olfactory, tactile and so on – and contributes to their semiotic formation, to their articulation in signifying substances. It is present on the plane of content of these very languages. It is imported within them, and constitutes that level of basic figurativity which makes abstract thematic content manifest, concrete and, therefore, credible. However, at the same time, it locates itself at the 'higher' level of its content, acting as a basis for the formation of semantic categories and their articulation. Openings and closures, dilations and contractions, impulses and resistance, energies and materials, forces and forms, euphoria and dysphoria, aims and grasps are equally prototypical corporeal processes that can be cognitively reinterpreted as states and transformations, actions and passions, operative programmes and modalities of doing, and perhaps, even earlier, as opposites, contradictions, complementarities, affirmations and negations.

The body also intervenes in an even more decisive way, at the moment of the enunciation of the discourse and traces of it can be found during the construction of various texts. The moment of enunciation is not an empty form that tends to produce (or to receive) discursivity. There is a body that, by taking up a position in space and time, determines itself and all of its own corporeal alterity in a different way each time. And it also inevitably constitutes its Me and its Self, its internal and its external, its being fickle flesh (internal and invisible) and its being an actual body (formed and manifest). We hypothesise then that the body is responsible not only for discursivity but also, on a deeper level, for semiosis, as it is from the corporeal assumption of a position that something is perceived as exteroceptive (an expression) and something else as interoceptive (a content).

V.6 Synaesthesia

What we now need are analysis models to connect sensoriality and textuality. On the one hand, this does not seem particularly taxing if we remember that textuality not only involves everything understood as a text by our language and culture, but also includes actual situations and experiences, along with practices, objects and, obviously, bodies. On the other hand, it is clear that sensoriality and corporeality cannot be reduced to canonical textual forms (stories, films, songs, adverts and advertising spots, etc.) within which they can place themselves on the plane of content. They require semiotic models of analysis that are both specific and general: specific because they are developed from the particular way in which the body and perception access meaning, elaborating and modifying substances and forms of expression; and general because they are in some way retroactive with regards to canonical textuality, which can also (in some ways) be reinterpreted *marks* left by the body of enunciation on certain materials, forming them through its own apparatus and producing substances that are different each time. And so the fundamental difference between the represented body (at the level of the enunciated) and the representing body (at the level of the enunciation) is very difficult to grasp. The case often arises of spoken (or depicted) bodies starting with speaking (or depicting) bodies, which in turn inscribe their marks into the enunciated, into the body they are somehow speaking about (or depicting).

Starting with these kinds of considerations, semiotics has attempted to develop a model for textual analysis relative to corporeality. It is a *somatic topic*, a matrix that establishes a typology for the modalities of sensitivity. Processes that are generally excluded from the perceptive apparatus but that are pertinent to a semiotics of the body, such as sensory-motor control or intimate bowel movements, are included here. This matrix starts from the assumption that the sensory body should be differentiated from the *body itself* (which produces a perception of the Self) and *flesh* (an example of intimate reference that determines a Me). Both these aspects of the body have an inside and an outside, as well as a border zone, that separates whilst also bringing what is inside into contact with what is outside, thanks to its varying levels of porosity. Sensory perception (which is the simultaneous perception of the world and the self) is a double process that goes from the outside in or from the inside out.

It is for this reason, for example, that the sense of smell, intimately bound to the breathing process, not only tends to bring inside that which lies outside the body, but is used in various cultures to semanticise life and death, birth and deterioration, in a very close relationship with the universe of the sacred. And it is for the same reason, as anthropologists know well, that in so-called primitive cultures the digestion and cooking processes (generally comparable in the field of taste) are considered one and the same: cooking is the cultural transformation of foods while digestion is the same thing carried out naturally.

A schema of this kind recognises the kinds of processes that can be found more or less transposed in various kinds of texts, be they descriptions or direct accounts of corporeal phenomena, or another kind of textual manifestation present in discursive fields of any kind, such as advertising, cinema, the arts and so on.

Most important from our point of view is that each kind of sensitivity (which does not adhere to the traditional division into five senses) installs a specific *figurative syntax*, a *formal articulation* of its processes that can be invested by different *substances*, giving rise to synaesthetic effects that are masked, creative, naturalised to different degrees. If a flavour can be visually, olfactorally and tactilely recommunicated (think of the complex lexicon used by oenology), if a series of sounds can be understood as a chromatic series (like in Rimbaud's famous poem titled "Voyelles") then it is because the modalities of taste and hearing each have

104 — V Image, sensoriality, body

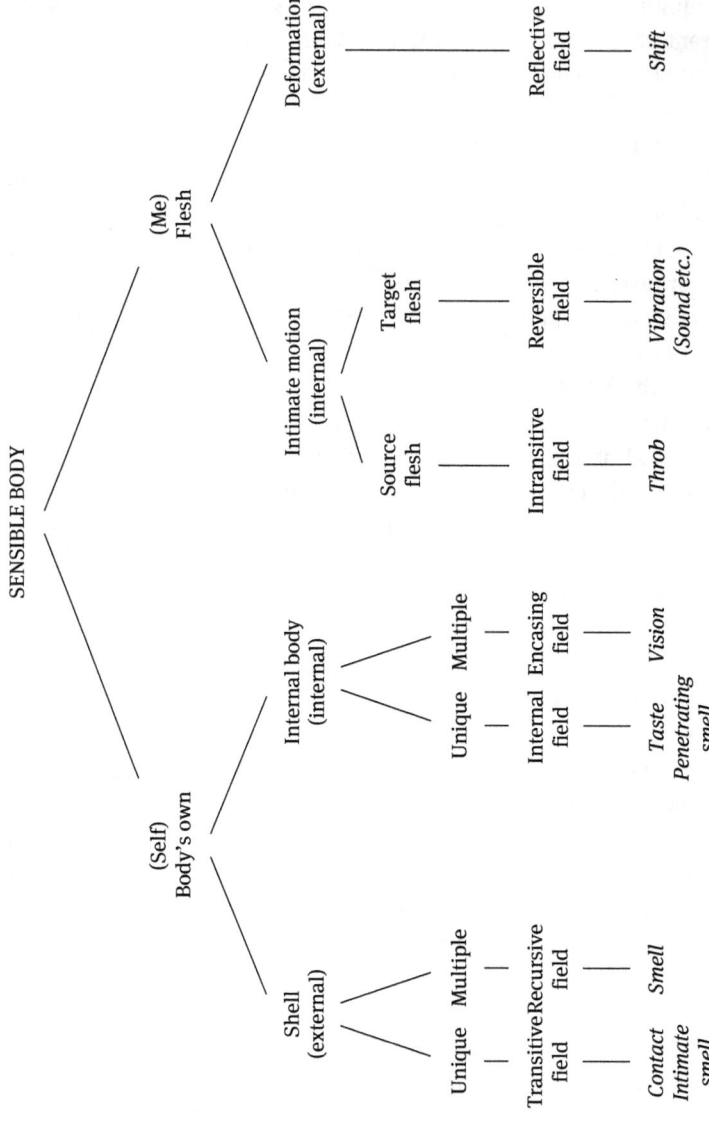

Fig. 19: Kinds of sensorial processes (from Fontanille 2004).

a figurative syntax that is invested at times by visual, olfactory and tactile substances, and at others by chromatic substances. We can see, then, that synaesthesia is not the external representation of one sense through another (like when we say that the image of a wet bottle gives the impression of a cold temperature), but the re-use of a syntactic form through other sensory substances. So, we have the re-emergence of that corporeal *common sense* from which the sensory modalities can differentiate themselves. Just as two images can be associated, giving rise to possible visual reasoning due to the fact they share a figurative sub-level, two sensory processes can, in the same way, be compared or overlap because they share an underlying syntactical articulation (Fontanille 2004).

Synaesthesia is often understood as a perceptive overlapping of expressive traits and semantic ones (the softness of a curved line → softness of an essence; the intensity of a colour → the intensity of a smell; a balanced form → a balanced bouquet, etc.). Moving from an analysis of elementary traits to an examination that takes into account entire discursive segments, starting with a formal underlying syntax, the range of possibilities is broadened and becomes more precise. For example, everything that in perfume advertising is usually read in terms of purely evoking a particular atmosphere (i.e. subtle allusions, a mood being delicately evoked, that ineffable '*je ne sais quoi*') can be understood in a new way. What is observed is not the figures themselves and much less what they might produce in terms of 'sensations' in the enunciate but their profound forms, their figural and plastic traits. These are the forms (common to visual and olfactory substance) that allow for the *textual translation between different sensorial languages*, making it possible to 'say' a perfume through an image, to transmit its meaning through visuality.

So, in order to explain the signifying articulation of a text where smells are concerned, the canonical syntax of smell must first be reconstructed. The olfactory processes require not only the body that perceives the smell (*target*) but also the body that emits it (*source*). A smell, whatever it may be, comes from an organic substance (natural or or synthetically reconstructed). Anything but spiritual, a smell brings into play two very different bodies that, entering in a relationship, bring together two kinds of syntaxes.

In terms of the body-target, the smell is manifested (progressively or suddenly) within a precise process (*emanation*) that follows three phases:
- *emission* (which deals with the smell's provenance);
- *diffusion* (the organisation of the space that is filled to varying degrees: the 'halo');
- *penetration* (in the double sense that it both enters into the body and the body enters into the halo).

These phases can either all be present, or some of them can be reconstructed through presupposition. So, if we find ourselves inside a smell, we must have first entered it; if the smell has spread out, it has come from a particular source.

In terms of the body-source, things are different: a smell requires a body that emits certain smells on the basis of its organic state, hence the *birth* of a smell, its *degradation*, its final *decomposition*. If a smell is fresh, it is because the emitting body somehow finds itself at the moment of the smell's birth; if it is stale, the decomposition process is in an advanced state. In any case, this is in some way the mark of organic conditions on the body-source. The interweaving of these two syntaxes gives rise to specific olfactory experiences that are sometimes very complex, and their relevant synaesthetic translations, providing us with the following schema:

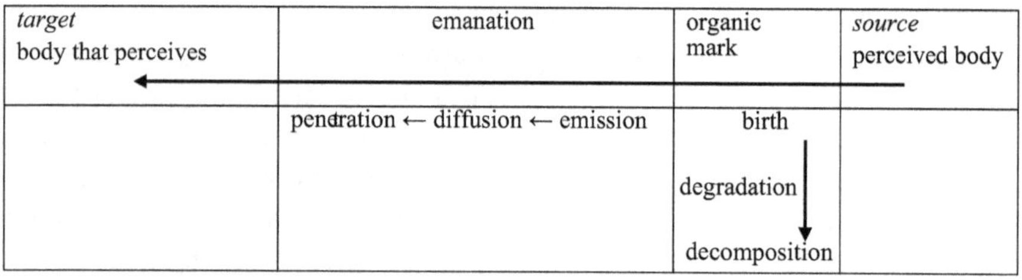

target body that perceives	emanation	organic mark	*source* perceived body
	penetration ← diffusion ← emission	birth degradation decomposition	

Fig. 20: The syntax of smell (elaboration from Fontanille 2004).

What is clear in these experiences is that questions are raised regarding space (the place of birth and emanation, the space of the halo's diffusion, the point of arrival, the distances between these two, internal paths and so on) in all of its possible juxtapositions (a 'musty' smell is one that has stayed in a place for a long time). The result is both a wide range of stories (seduction is a continual negotiation between emanation and penetration) and, most importantly, an enormous number of visual depictions starting from the interplay between the inside and the outside (the penetration of the body by a smell/ the immersion of a body in the smell; envelopment in concentric circles; being startled by sudden emanations, etc.). Just think of the abundance of perfume advertising, stuffed full of figurative elements such as clouds, bright lights, rays of light, windswept hair, water and various other liquids, flowing fabric, petals, and so on. These are transpositions of halos, emissions, penetrations, diffusions and emanations onto the figural plane.

V.7 Space and subjectivity

As we have said, issues of the body are not linked exclusively to the sphere of the senses. They involve many other semiotic phenomena, including space. A perspective on the world and part of the world, but also the porous casing equipped with an outside and inside which are redefined on the basis of varied multi-sensory experiences, the body has a constitutive nexus with spatiality. It acquires consistency and meaning through its placement within a determined place, in the space it experiences, lives; and at the same time the body helps to provide these places with meaning, to constitute their physiognomy and signification. Understood semiotically, the space and the subject construct one another through the body with its double nature of an element situated within a space and a spatial extension, both contained and containing.

From the perspective of the analysis of human and social signification, space is a system and process of signification, a structured collection of things and voids that speak of things other than themselves. Spatiality speaks to society about society and its articulations, its institutions, its transformations. Just as verbal languages are articulations of sounds in reciprocal presupposition with articulations of meaning, spatiality is a series of structured extensions (real or metaphorical) in reciprocal presupposition with a series of structures of signification that are culturally and historically determined. There is no reason, then, for the traditional conceptual opposition between objective and subjective space, nor for that between functional and symbolic space. For semiotics, *spatiality and subjectivity constitute one another reciprocally*. Semiotics understands spatiality not as a physical or natural environment but as a signifying phenomenon for humans, a space that human cultures constitute as their own tool for signifying the social universe. And the subjectivity is understood not as individuality or a single conscience but as a something that is both pre-subjective (corporeal) and inter-subjective (social). Subjectivity is not pre-constituted or static; rather, it is constructed through processes that involve both perception and the body, both the sphere of sociality and that of culture.

So, the semiotic perspective, studying urban spaces (places and cities, shops and apartments, shopping centres and offices, and so on) focuses on the user's point of view, the body-subject who traverses these spaces, submitting to them or transforming them. Also, because of the individual and collective re-appropriation of a place, the real way in which it is experienced, understood, valorised lies between urban planning's predetermination of a place and its signified. As urban designers and architects, interior designers and city planners, geographers and administrators are all well aware, the end goal is to predict the social destiny of a space, to guide people's behaviour within particular environments that are more or less vast, more or less institutional. However, as they go about their lives these people meet one another, pursue common objectives or enter into competition with one another, they re-semanticise those environments, adapting them to their own needs, re-purposing them for their own actions and, most importantly, passions.

In developing this kind of theoretical perspective, the semiotics of space insists on a number of related issues. The first is narrativity: space signifies when, by articulating itself, it inscribes within it the series of actions of those who inhabit and experience it. This series is *narrative*: the actions are significant because they are articulated in view of NPs that are complex to varying degrees. If spaces are functional for someone, it is not in an instrumental sense but because they encourage (or impede) certain forms of behaviour. Inscribed within the shape of an apartment is the style of life of the person who lives there; the articulation of a city is the reflection of the relationships between institutions and citizens, individual and collective subjects. In this sense, the urban space should not be understood as a container but as a series of actants that do (or make-do) and provoke passions in other actants, according to the polemical-contractual relationships found in classical models of narrative grammar and the strategic games that are determined within every circumstance.

Detailed analyses of urban spaces have used the canonical narrative schema, observing how beneath traditional places (buildings, roads, pavements, benches, etc.), precise portions of actantial space are constructed in inhabited urban areas. These portions of space can be linked to manoeuvres of manipulation (meeting places around which to make evening plans), or to the acquisition of skills (places where one is allowed to practice sports), or the staging of a performance (various forms of entertainment and pastimes, challenges between groups, nocturnal escapades) or the enactment of a sanction (be it negative or positive). We could reason in a similar way with the canonical pathemic path, which only partially overlaps with the narrative schema. Doing so leads us to rediscover places of constitution (where affection without a name emerges: clearings, open spaces, boundless horizons) and pathemisation (where passion leads to action: acknowledgments, pushes, accelerations), places of emotion (where passion takes over the body: shouts, jolts, passionate embraces) and moralisation (where everything becomes either vice or virtue: punitive closures, euphoric openings) and so on.

We also know from narrative theory that the programme of action of a certain subjectivity opposes the programme of another. If a certain space appears to be significant, it is because a narrative polemic, a series of strategies or intersubjective tactics, are inscribed within it. An obstacle is an impediment to entry, while access is an invitation to enter. In every space there are at least two subjects that enter into a relationship with one another, actants equipped with the modalities of doing and the resistance to these actions. It has been demonstrated how the shape of an office desk favours specific proxemic relationships between those subjects who enter into a relationship with one another, consolidating or nuancing the hierarchies between management and staff. There are desks for *decision-makers* (which produce the effect of a single, uni-directional space that solicits the cognitive dimension), and those for *pilots* (whose space is multiple and multi-directional, functional for the pragmatic dimensions); there are desks that deny the practical activity of the boss (producing a re-centered and re-directed space around which people meet) and those that deny cognitive activity (producing a bi-directional space in which one engages in discussion with others). And this is not behaviourism: inter-subjective relationships are not inscribed in space only in the sense that means certain places provoke certain kinds of behaviour, but in the sense that, more profoundly, such places *do* in place of the subject, assuming human actions and making them their own.

Spatial organisations are non-human actors that are a little better at doing what a human actor could do: a lift or an escalator climb the stairs for us; an automatic gate substitutes a porter; an open door invites you to enter, a closed door denies entry. When we come into contact with a well-warmed room, we are not only experiencing a state of the world, as that state is founded on a series of individual and collective actions that have allowed the room to be warmed. To say all this using semiotic terminology, every physical place is an utterance of Being that presupposes at least one utterance of Doing. And in order to reconstruct the meaning of that place, it is necessary to reconstruct the chain of presuppositions, the series of mandates, and rediscover the human operations behind the things of the world that have equipped them with meaning.

Thus, it becomes possible to distinguish between:
- *enunciated subjects in space* (whereby certain actions and programmes are delegated to spatial forms and substances, as with an elevator);
- *subjects given at the level of enunciation* ('model users' provided by spatial articulation);
- *empirical subjects that inhabit the space* (adapting themselves to model users, accepting the semantics of the space in which they find themselves, or alternatively, re-semanticising it).

The first are implicit and physical, often more efficient. The second are abstract and bound in some way by the situation. The third are tangible and unpredictable. But more important than the typology itself is seeing how it is possible, in practice, to pass from one of these subjects to the other, observing not only how the pre-constituted and abstract model directly influences people, but also highlighting how the empirical can influence the model by re-semanticising the space (think of airports and train stations where the traveller/ customer not only goes to catch an aeroplane or a train, but who also uses those places for meetings, to stop at the restaurant, to buy a book, etc.) (Marrone 2001).

One example of this kind of phenomenon is provided by the way in which spatial articulations of shopping centres, supermarkets and hypermarkets predetermine the typology of their clients, configuring precise forms of consumption that are at times careful and at times distracted, at times functional and at others, playful. There exists a kind of reciprocal presupposition (a relationship that is not casual but one of signification) between the forms of articulation of the space (continuous/ discontinuous, non-continuous/non-discontinuous) and the forms of consumer behaviour (Floch 2001). Long corridors, all relatively narrow, ensure that the people with the trollies channel themselves into anonymous trajectories, one after the other, awaiting their turn in front of the displays and grabbing products only when they arrive within reach (acting like *sleepwalkers*, customers lacking their own initiative). Conversely, wide, complex spaces, with no immediately perceived order, lead consumers to make their own routes, reinventing the way shopping is carried out, the products chosen, the meaning given to it all (behaving like *explorers*, characters that are active and aware of their own actions). Similarly, non-discontinuous spaces, filled with fractures, limits, jumps, lead people to pre-organise genuine buying sequences, selecting in advance exactly where to go and what to do (giving rise to forms of shopping *professionalism* that are in no way creative or open to novelty). Finally, non-continuous spaces that are open and do not feature pre-established directions, with multiple entrances and exits, lead the consumer to relax to the point of almost forgetting the reason they are there, or rather, to the point of modifying the very reason they are there: not to shop but to walk around (like *dawdlers* who, with no given programme, go to the place to decide what to do that day). This gives us the following framework:

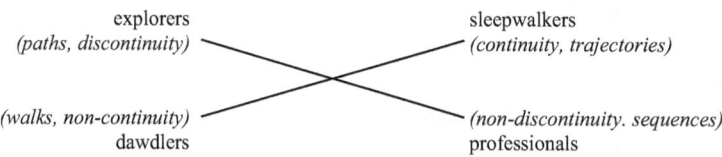

Fig. 21: The semiotic square of subway passengers in Paris (from Floch 2001).

It follows that the sphere of affectivity and passions should be inserted into a consideration of spatiality. Space signifies also because it provokes passions, or is itself impassioned. It has its own passions instead of subjects that inhabit it, it undergoes actions and suffers. For example, if a *want to see* of a living subject that lives in a certain place corresponds to a *want to not be seen* of that place, this sets in motion a kind of voyeurism that provokes pathemic reactions such as discomfort, frustration or similar. In this sense, a glass wall is a kind of architectural 'promise' since it allows us to see what it will be possible to physically reach, only later, once all the obstacles have been overcome. And, like all promises, it can be kept or not, giving rise to a series of actions and passions, to stories that can be very complex.

From here comes the possibility, for example, of a *semiotics of shop windows* (Oliveira 1997). For the science of signification, shop windows are not simply thresholds but are machines producing subjective and inter-objective meaning, dispositives of urban décor, processes of seducing or luring the customer. With the shop window, one exposes oneself, is rendered entirely transparent, or locks oneself up. The outside anticipates that which will display itself more clearly inside, as well as simulated realities being set in motion, with objects that are more or less incongruous placed in relation with one another. There are modest window displays and immodest ones, discrete windows and 'pornographic' ones, with the entire corresponding series of possible consumer reactions that are inscribed in the material and spatial dispositive of transparency: I can accept the invitation to look and to enter, but I can also refuse it; I can glimpse inside, I can feel like I am already inside, I can perceive a resistance, I can even see my reflection.

V.8 The aesthetic grasp

Lastly, we come to the problem of the relationship between sensory and somatic processes that we have referred to as aesthetic (or aesthesic, if we are being strict), from the Greek *aesthesis*, and all those other phenomena that seem to be of an entirely different order and that fall within the domain of art, beauty or, as it is sometimes rightly known, the aesthetic experience. Scores of thinkers, scholars and critics have questioned themselves on the nexus between sensoriality and art, beauty and corporeality, and this is not the place to revisit their responses, which vary in order, importance and complexity. From the point of view of a textual semiotics, which is the critical analysis of every social process and cultural phenomenon, it is necessary to place to one side the problem of attributing artistic value to some texts (so-called works of art), an issue which is both philosophical and sociological. These vary enormously between cultures and historical periods, according to the criteria used each time to define them. Rather, it is necessary to study the ways in which sensory and somatic processes are textualized, inserted into some configuration of meaning and, as such, are equipped with a specific signification that, in some cases, is overdetermined as artistic. And it is necessary to examine if and how these processes that textualise sensoriality and the body function in some of those texts considered works of art, to the point that they can be viewed as discursive models.

From here comes the notion of *aesthetic grasp*, useful for dealing with this kind of textual problem and, theoretically, for eliminating various clichés regarding the nature of body and sensory processes. From a semiotic point of view, the body is not the material basis of the individual conscience. It is something that precedes and, at the same time, transcends individuality. It is pre-subjective and inter-subjective, it smashes the conscience to pieces and dissolves it in the socio-cultural sphere. Similarly, the senses are not the starting point for the cognitive relationship between subject and object, as dictated by philosophy. These emerge only if it is possible to place to one side those perceptive grids that, through specific cultural models, direct that very perception cognitively and pragmatically, giving the senses new expressive potential. The aesthetic grasp is that unsayable moment (in the sense that it cannot be predicated through established linguistic and cultural codes) in which sensoriality re-emerges, unaware but determined. It blooms 'after' like something that should have been 'before', radically transforming cognition, subjectivity and inter-subjectivity.

What does this involve? In order to explain it we must remember two points. The first is that of the presence of figurativity on the plane of the content of languages so that perceptive processes contribute to the construction of the signified. Meaning is not just conceptual, but is also sensitive, because sensitivity is not only produced on the plane of signifying expression, but also on that of signified content. This is why a focus on sensitivity and the body is not so much necessary for outlining the specificity of languages starting from the material used on the plane of expression (as with semiology in the 1960s) but, instead, implies a focus on semantics. The second point is the relationship between plastic and figurative: the plastic, on which much visual signification is founded, emerges when the figurative apperception of the image is transcended. The plastic is something perceived in a later moment, when the figures of the world are placed to one side (perceptible because they are already signifying for us), and which provides a new formation of the expressive substance. So, no more figures, but graphic traits, contrasts of colour, position and dimensions of the masses in the format. The plastic is a kind of 'other' vision of the world which requires uncommon abilities that are not immediately provided socially, very often the prerogative of artists or, in no way paradoxically, of the stupid. This is the reason why, as is often said, artists have a particular sensitivity: they see the plastic where everyone else perceives the figurative. It is just like the stupid person who, when they take the Rorschach test, sees not figures of the world in those marks but mere shapeless stains smudged over the page. The plastic is not the 'pure' emergence of the substances of the world, but their different formation: new, unforeseen, unpredictable. This is the reason that it is often considered the source of artistry, if not beauty. It is something that stands, as we are well aware, not only for vision but for any sensory process that goes beyond common perceptive grids, allowing other dimensions of meaning to emerge in smell, taste, hearing and so on.

While this particular ability, which we could refer to as *plastic skills*, is constitutive (as it were) when it comes to artists or the stupid, it is nevertheless present to varying degrees in all of us, and contributes to the constitution of subjectivity which, as narrative grammar teaches us, is also the result of a transformation. The aesthetic grasp is therefore the *non-narrative transformation of the experience*, the corporeal constitution of subjectivity. The subject institutes itself in the various phases of the tale thanks to the logic of their programmes of action and passion. Yet, on a sensory level, subjectivity can transform if it finds itself set in those unpredictable moments in which perception reforms the substances of the world, often becoming passive before a 'beyond' that, actually, acquires active roles. So, when it then returns to a normal perception of things, it is different, changed, even if it is impossible to say quite why or what actually happened, leaving an aftertaste of imperfection and a sense of nostalgia.

These kinds of phenomena, explained by psychologists, philosophers and anthropologists alike, can be analysed on the level of semiotic structures in cultural texts. Semiotics searches its texts for those processes in which the body is a complex collection of sensoriality and perception, but also of programmes of action, cognition, passion, and so on, in order to see how those texts are textualized, giving rise to discursive models that can sometimes be generalised. In other words, for semiotics, texts are those configurations of meaning in which a series of phenomena, including those that are corporeal, acquire (or change) a certain role, taking on some signified. These can be texts of any kind, from literary ones to those from everyday life, produced spontaneously (as it were) by culture, with no artifice or ulterior motive if not that of producing and circulating human and social meaning. In this way, sensoriality is not studied thanks to specific interpersonal sets or laboratory experiments, but emerges directly in the ways in which it is presented culturally and textually.

V.9 A final glance

In order to illustrate these problems with an example, and to indulge in a small exercise of textual analysis with which to bring our book to a close, we will use a famous story by the Italian writer Italo Calvino (2002: 8–11) from the collection,

Mr Palomar, entitled "The Naked Bosom". This is the part of the book where Calvino describes the troublesome encounter between the 'philosophical' protagonist of the book (named Palomar, after the astronomical observatory in California) and a woman sunning herself topless on the beach. Faced with this insignificant yet embarrassing event, Mr Palomar poses himself various ethical and ideological problems: should I look at the woman? And if so, in what way – by including her in the landscape like any object of the world or, conversely, by paying discrete but satisfied attention to those "haloed cusps" exposed with probable mischievousness to curious passers-by? One of the gazes chosen by Palomar with which to look at the woman is a good example of the aesthetic grasp. The text recounts the process by which vision is converted into touch; the gaze finds itself brushing against the woman's bosom, who ends up leaving, irritated. This event, seemingly empty of any importance, brings about a profound transformation in both Palomar and the woman: both lose their own identity, they become pure gaze and pure bosom respectively, before taking on another identity. Let us look at what the text says in more detail.

Mr Palomar is presented in Calvino's book as a kind of positive hero whose knowledge is denied. By practicing a cognitive minimalism, Palomar refuses to totalise the experience of the world around him by enclosing things and events in systems of concepts, preferring to concentrate on punctual descriptions of what he sees. Palomar puts into practice the phenomenological precept by which reality is constructed through the progressive imaginary elaboration of successive points of view on it. He starts from an observation or sensation, and slowly transforms each, developing their possible facets. Yet, at the moment of truth, when it comes to making a decision, he misses the numerous, often contradictory possibilities that he is offered, preferring to close himself within in an obstinate silence, doubtful but knowing. Calvino writes, "Rather than the cultivation of precise nomenclature and classification, Palomar had preferred the constant pursuit of a precision in defining the modulating, the shifting, the composite". He, like many artists, tries to define the undefinable, a place where the knowing subject discovers the vainglorious result of a previous repression: as Palomar says at the end of his visual wanderings: "the 'I', the ego, is simply the window through which the world looks at the world."

So Palomar oscillates between four different ways of looking at the woman's bosom, and four ways of re-elaborating it cognitively. Every somatic action (walking, moving his head, directing his gaze) is followed by a cognitive action (reflection, doubt, adherence to cultural stereotypes), at a gathering pace that only the woman's final escape manages to halt. Every gaze is the result of an ethical assumption, but every ethical assumption is the result of the visual *performance* that precedes it. Hence the four *forms of life* generated by the interweaving of the aesthesic dimension and the relative models of behaviour.

The first moment is marked by durativity: a soft, partial adhesion to a common sense that carries a generic ethic of discretion. Upon encountering a topless woman among the few bathers, "Palomar, discreet by nature, looks away at the horizon of the sea". He, says the text, "knows that in such circumstances, at the approach of a strange man, women often cover themselves hastily, and this does not seem right to him". The stereotypical aesthetic judgement, albeit issued in the negative, is the immediate consequence of a shared and unaware knowledge to which, up until this moment, Palomar adheres. The text underlines how the gaze directed at the horizon is iterative, meaning it stands for the numerous times in which our hero has found himself in similar situations: "And so, as soon as he sees in the distance the outline of the bronze-pink cloud of a naked female torso, he quickly turns his head in such a way that the trajectory of his gaze remains suspended in the void and guarantees his civil respect for the invisible frontier that surrounds people". If the indefinite article ("*a* naked female torso"), in contrast to the definite article used in the title ("*the* naked bosom") harks back to a typical moment of life on the beach, the urgency that follows ("quickly") suggests a kind of automatic, standardised response, the result of thoughtless habit and ethics. The "invisible frontier", on the other hand, will be precisely what the ensuing plot will attempt to breach.

However, as soon as "the horizon is clear" and his "eyeballs" recover their full range of movement, Palomar substitutes common sense with his own personal reflection. He understands that his is not a way of looking that presupposes nudity, with all of the taboos this entails. So he passes once more in front of the bather and

> keeps his eyes fixed straight ahead, so that his gaze touches with impartial uniformity the foam of the retreating waves, the boats pulled up on shore, the great bath towel spread out on the sand, the swelling moon of lighter skin with the dark halo of the nipple, the outline of the coast in the haze, grey against the sky.

Whereas earlier his gaze was turned to the horizon or remained suspended in mid-air, and kept in check by a responsible and proper subject, Palomar now allows his vision a certain degree of autonomy. Thus, the gaze synaesthetically transforms into touch: "his gaze touches with impartial uniformity" the collection of objects that bring the landscape to life, including, in the form of a chromatic detail, the woman's bosom. The gaze has already found its tactile dimension, useful for encountering the object. However, the latter, presenting itself as a pure object of vision, does not yet declare itself ready for the encounter. It is still in a chromatic dimension, neither tactile nor eidetic.

This leads us to the third kind of vision, predisposed by a further reflection: by inserting the bosom among the other things of the world, Palomar thinks, I will treat it like a thing, and by doing so I reiterate the sexist prejudice that reduces woman to a mere object. Here is the central fragment of the text:

> He turns and retraces his steps. Now, in allowing his gaze to run over the beach with neutral objectivity, he arranges it so that, once the woman's bosom enters his field of vision, a break is noticeable, a shift, almost a darting glance. That glance goes on to graze the taut skin, withdraws, as if appreciating with a slight start the different consistency of the view and the special value it acquires, and for a moment the glance hovers in mid-air, making a curve that accompanies the swell of the breast from a certain distance, elusively but also protectively, and then runs on as if nothing had happened.

Here, we no longer have Palomar and the bather, the man and the woman, the subject and the object, but only a gaze and a bosom. Having lost their social identity and their usual configuration, the two new actants in the story advance towards one another. On the one side, before the "neutral objectivity" of the surveying gaze, the bosom literally enters into the field of vision, activating itself as a subject before the two Addressers watching the scene from the outside ("he arranges it so that [. . .] is noticeable"). On the other, the gaze once again abandons its visual origin and, confirming the synaesthesia that has already been set in motion, acts like touch. The gaze advances, no longer grazing the naked bosom, and much less with the woman lying in the sun, but with the taut skin only, that sensitive quality of the world better suited to be perceived by touch. This is the moment of the aesthetic event takes place: the manifestation of a discontinuity not only in the field of vision but in that of inter-subjective existence. The moment in which the sight descends towards the lowest level of perception, and the materiality of the world offers itself as the sentient body, the two actants can finally conjoin in a moment of unexpected happiness. The "darting glance" (in Italian: "guizzo") evoked like a metaphor for the visual offcut, containing within it the sudden movement and the quick glow of light, are the words that provide figurative synthesis of the aesthetic grasp.

The theme of the restless gaze is prolonged in the moment that follows immediately after, involving the aesthetic appreciation of conjoining, which is also tactile and not cognitive. The "slight start" is a jolt, a movement of the body caused by a sudden emotion. And while this jolt is attributed to the object, the emotion is felt by the subject: by uniting passions of the body with passion of the soul, the maintenance of actantial syncretism prolongs the moment of happiness for just while longer. His gaze momentarily "hovers in mid-air" before losing its tactile grip and returns to that plastic level of visuality that is the eidetic dimension: tracing a curve, the eye gives the bosom back its abstract shape, translating the tactile effect into the order of the plastic. It continues "as if nothing had happened".

To summarise, here are the main moments of the scene in order:
- Palomar arranges things so that the discontinuity is noticeable;
- the bosom enters into the field of vision;
- the gaze advances;
- the bosom becomes taut skin;
- the vision becomes touch;
- the conjunction happens ("darting glance")
- the slight start;
- the gaze that hovers in mid-air;
- the gaze returns to the eidos;
- the bosom is reconstituted as a shape;
- the gaze continues along its way.

We are left with a number of questions: what does "as if nothing happened" mean? Why does the woman jump up and move away "as if she were avoiding the tiresome insistence of a satyr"? What has happened? In order to see the problem, it is necessary to read what happens next. If we pay careful attention to the reflection with which Palomar follows this visual performance, we notice that he is conceptualising not so much the actual aesthetic event, but the moment that follows it: the one in which his gaze, albeit in its eidetic depths, always only remains in the visual dimension. The text states that "this grazing of his eyes" could be interpreted by others as a desire to place the bosom itself to one side, undervaluing it and relegating it to the usual, centuries-old "sexomaniacal puritanism". In short, the cognitive rearrangement of the aesthesic ends up denying it its own truth. If Palomar's gesture suggests "puritanism", it is because he is silent about the fact that the conjunction of the two subjects has occurred, albeit it in an aesthesic way: "as if nothing had happened" conveys precisely this ambivalence, this two-fold way of reading the event: on the sensory-aesthetic plane and on the more traditional, cognitive plane.

This brings us to Palomar's fourth and final attempt, which nevertheless remains entirely virtual. Now he wants to express a "detached encouragement" to his gaze in the direction of the change in habits, his predisposition to accept new social stereotypes that remove "the idea of amorous intimacy" (or rather, exactly what has just happened) from the female bosom in the name of a "broad-minded society". But the recovery of the dimension of shared knowledge ends up leading to a return to the figures and icons of the world, in a tactile dimension that is merely hinted at:

> Now his gaze, giving the landscape a fickle glance, will linger on the breast with special consideration, but will quickly include it in an impulse of good will and gratitude for the whole, for the sun and the sky, for the bent pines and the dune and the beach and the rocks and the clouds and the seaweed, for the cosmos that rotates around those haloed cusps.

As we can see, the gaze no longer grazes anything, instead limiting itself to granting a "fickle glance" to the landscape that includes the bosom among its other elements, a bosom re-considered in the fullness of its being, like a figure of the world that is already conceptualised and culturalised. The subject's "impulse" gives way to the rotational movement of the cosmos, losing any desire for conjunction. The action, furthermore, is entirely in the future, and Palomar's cosmological proposal will never take place: "the moment he approaches again" the woman gets up and leaves. We can understand why: the woman, like Palomar, has not cognitively processed the aesthetic event, and she flees the comings and goings of a time-waster. Or, more convincingly, she has processed it perfectly, and she really is fleeing the satyr who is rushing to touch her once more. Essentially, in the field of vision there was only her bosom, not her as a whole, body and mind. In one way or another, the stereotype has overwhelmed Mr Palomar's "most enlightened intentions", leaving him with a vague nostalgia about what has happened and a clear sense of human and social imperfection.

Recommended bibliography

On visuality, images and iconic text: Barthes 1977; Basso-Fossali & Dondero 2011 (on photography); Bordron 2011; Calabrese 1985 (an introduction to semiotics of arts); Corrain 2016; Damisch 1972, 1984, 2012; Didi-Huberman 2005; Dondero & Fontanille 2012 (on scientific images); Dondero & Novello-Paglianti (eds.) 2006; Elkins 1999; Fabbri 2019 (on contemporary arts); Floch 1997, 2000; Florenskij 2002; Fontanille 1995 (on lights and colors); Freedberg 1989; Gell 1998; Gombrich 1951, 1960; Latour e Weibel (eds.) 2002 (on the concept of iconoclash); μ Group 1991 (on the rhetoric of images); Mangano 2018 (on photography); Marin 1995, 2002; Migliore 2021 (on the reception of images); Ouellet 1992, 2000; Panofsky 1955 (on the levels of meaning in pictures); Saint-Martin 1990; Shapiro 1973; Stoichita 1989 (on the origin of the picture), 1997a, 1997b; Thürlemann 1982 (on Klee); Wölfflin 1915 (on the basic priciples of the arts).

Specifically on the relation between figurative and plastic languages: Calabrese 1999, Corrain 1996; Floch 1995; Greimas 1984; Patte 1990; Marrone 2007; Dondero, Beyart-Geslin & Moutat (eds.) 2017; Mangano 2018; Dondero 2020, Zinna 2019.

On space and spatiality: Augé 1995a; Bachelard 2014; Bertrand 1985 (on spatiality in the novel); Brandt 2004; Cavicchioli 2004; Giannitrapani 2013 (an introduction to the semiotics of space); Hall 1966 (on proxemics); Hammad 2002, 2006, 2015 (fundamentals

in semiotics of space); Jaworski & Thurlow (eds.) 2010; Marin 1973 (on utopic spaces); Sedda & Sorrentino (eds.) 2020 (on islands); Zunzunegui 2003 (on museums).
In particular, on urban spaces: Marrone (ed.) 2010; Leone (ed.) 2008; Pezzini (ed.) 2009; 2016; Sedda & Sorrentino 2019.
On sensoriality, perception and aesthetics: Beyart-Geslin & Novello-Paglianti (eds.) 2005; Broden 2009; Landowski & Oliveira (eds.) 1995; Landowski, Dorra & Oliveira (eds.) 1999; Oliveira (ed.) 2013, 2014; Parret 1988, 1991.
On aesthetic grasp: Greimas 1987a; Landowski 2004; Landowski, Dorra & Oliveira (eds.) 1999; Marrone 2017b, 2019.
On the body: Finol 2021.

Appendix: A History of the notion of text

> The text, in its mass, is comparable to a sky, at once flat and smooth, without edges and without landmarks; like the soothsayer drawing on it with the tip of his staff an imaginary rectangle wherein to consult, according to certain principles, the flight of birds, the commentator traces through the text certain zones of reading, in order to observe the migration of meanings, the outcropping of codes, the passage of citations. (Barthes 1975b: 9)

1 Text and society

> A science which studies the life of signs as part of social life.

When, in the early 20th century, Ferdinand de Saussure (1988) was looking for a new scientific object of linguistic knowledge, he proposed this definition for the *semiology* or *science of signs* that would play a very important, though controversial, role in the history of culture from then on. What strikes the reader about this definition, more than the appeal to science, is the double reference to *life*: of signs and of society at the same time. That is probably because, as the brilliant linguist must have thought, they are basically the same thing. Though many later authors severely criticized structuralism accusing it of having an abstract nature, of being full of formalisms and closed to the outside world, here, at the very beginning of our history, we can see instead the *dynamic signs of dynamic society* in need of a science to explain how they work and to understand what their rationale is. Languages, discourses, and signs are social processes: their necessary and fundamental formal nature cannot but point at the fact, by underlining, confirming and proving it.

What happened after that? At first, since authors such as Roland Barthes and Umberto Eco adopted and developed the Saussurean concept, the focus on sociality was essential. At that time, the rising mass culture, with its peculiar communication media, the emerging consumer society, design, but also literary and artistic experimentation, the renewed logical-linguistic attention of philosophy and the development of an autonomous epistemology of human sciences, brought with them the need for a careful and disillusioned theoretical perspective that would be able to set up a *formal method of analysis of society*, free of any underlying ideology. Semiotics – as Saussure's chimera came to be called, sometimes in preference to semiology – addressed that need, and right at the half of the century it was born as an individual science, with its own authors and its own institutions. Books like *Mythologies* (Barthes 1957) and *Apocalypse Postponed* (Eco 1964) testify to the attention of the science of signs toward everyday social life, and the vocation to *criticism* – in every sense of the word – that such a perspective could not lack, focused as it was on systems and processes of signification. In order to study television or advertising, popular songs or comics, journalism or fashion, artistic avant-gardes or experimental novels, food mythologies or wrestling matches, it is necessary to gradually learn to look at them, combining linguistic competences and sociological interest, methodological attention and critical sensibility, formal vocation and a philosophical in-depth perspective.

However, the semiotics of the following period has mainly abandoned that original trend. The science of signification of the European tradition choose to walk other ways. The only exception was the Anglo-Saxon tradition and its *social semiotics* that, however, fell into the wider field of *cultural studies* (thus losing its methodological accuracy but gaining a controversial twist). The European science of signification focused, on the one hand, on a kind of semiotics that would investigate culture, sharing its research field with human sciences such as folklore, ethnology, comparative linguistics, religion studies, historiography, psychoanalysis and sociology. This kind of semiotics would aim at building general models allowing an accurate study of anthropological mechanisms. On the other hand, it focused on an analysis of non-verbal languages – such as images, gestures, audio-visual material, everyday-life objects – that would propose accurate methods of analysis of any possible work of expression and communication, with the same success as structural linguistics.

Authors such as Algirdas J. Greimas and Jurij M. Lotman were able to walk both ways, and to go from the creation of a general cultural model to the accurate analysis of a single work. Thus, semiotic model allowed cultural anthropology to provide fodder for literary and artistic criticism, they allowed philology and iconology to endow themselves with an ethnological aspect and media studies to use linguistic methodologies and vice versa. At that time, however, the critical verve of the formal analysis of social facts got lost and so did the hypothesis that any linguistic, communicative, expressive or semiotic phenomenon had a social basis, as Saussure, prophet of pure differences, had seen so clearly. This kind of analysis focused on particular works (such as novels, short stories, poetry, movies, paintings, pictures, ballet, advertisements, TV programs, newspaper articles, architectonic works, objects), and anything that in our culture could be called a 'text' entered the field of investigation for semiotics. The 'text' is here regarded as *any expression medium able to convey certain meanings*, with specific characteristics and clear boundaries, with its own processuality and so on.

The science of signification has gradually expanded the notion of text and used it as a tool to study not just semiotic entities using non-verbal expression-substances, but also different kinds of cultural phenomena that can have the same basic properties as a book-text (biplanarity, closing, stratification of levels, processuality, etc.) without its precise print bearing. Thus even what is not regarded as a text *from an empirical point of view* is nonetheless analysed as such *from a methodological point of view*, for the reason that the same formal properties of texts can also be found in it. Such an analysis can be performed, for example, on TV show schedules, advertising campaigns, information streams, communication platforms, oral conversations, web interactions, marketing strategies, subway stations, buildings, even whole cities. In this perspective, *the text is not a thing* anymore – it is not an empirical object *but a theoretical model acting as a tool for description* that, given certain requisites and certain explicit epistemological conditions, is able to retrace, at different levels, the formal devices of any object of knowledge of the science of signification. As the concept of narrativity has been reached by gradually expanding the analysis of actual narratives (e.g. fairy tales, myths, novels, short stories and many other literary works) in order to explain seemingly non-narrative discourses (e.g. advertising, political, journalistic or philosophical discourses), so the concept of textuality has been built using actual texts (e.g. novels, poetry, pictures) as a tool to explain the structure of meaning in seemingly non-textual semiotic manifestations (e.g. hypermarkets, methods of cooking, scientific experiments and so on). The text is *the formal model that can explain all human and social, cultural and historical phenomena*. That is the reason why Greimas used to repeat 'Outside the text no salvation!' and why many semioticians still refer to their specific object of study as 'text' whatever its nature is.

Hence the new wave of *sociosemiotics*, proposed by Paolo Fabbri (2018 [1973]) as a methodological perspective for the study of all sociological phenomena. It was on this base that authors such as Jean-Marie Floch (2000, 2001), Eric Landowski (1989, 1997, 2004) and many others set out to explore a semiotic study of social facts, such as advertising, political and journalistic communication, fashion, design, cookery, public spaces, everyday life, objects, that was focused on their social and cultural values and their discursive implications. Since sociosemiotics did not analyse only given works, but something less well determined such as situations, practices, habits, sensorial and corporeal experiences, flows of information or media interactivity, the dichotomy between 'text' and 'context' had no reason to exist: many so-called contextual facts (e.g. those 'outside' of the given work's text), can also have a semiotic pregnancy of their own, within a coherent description project. While the linguistic perspective, even on its pragmatic side, differentiated linguistic and non-linguistic phenomena, according to sociosemiotics this difference is not given *a priori*, since both contexts and situations could be at the same time meaningful and social, communicative and factual, textual and experiential. From the sociosemiotic point of view we can call 'context' only what is not relevant to the analysis anymore; and it is actually culture that determines if something is or is not relevant, even before the analysis itself. The text, on the other hand, is no longer the material base for interpretations that could integrate it, or even justify its existence, but the *formal device through which the meaning builds up its structure* and thus shows itself, the device through which the meaning spreads in society and culture. It is therefore *the specific object of study for the semiotician* who analyses it and tries to retrace forms and dynamics, internal structures, levels of relevance, inputs and outputs. *The text is not a given entity*, nor a phenomenal evidence, *it is the result of a double construction: a socio-cultural configuration* before, and *an analytic re-configuration* afterwards. The text, in this perspective, is necessarily *negotiated* within cultural dynamics that, in creating the text, come into existence and interlace in an unending chain with other texts, other matters, other languages. *The text is not closed in itself but easily remodelled and configured in other textual forms*, easily *translated into other languages* in the never-ending inter-textual, inter-discursive chain of the semiosphere.

2 Sociosemiotics without textuality?

However, sociosemiotics nowadays occupies a more radical position and its original form has thereby vanished. The radicalisation of its position occurred because, according to a certain formulaic sociologism, it opposed, in a somewhat grotesque way, the study of 'texts' to the study of what is known as 'practices' (see Landowski 2004; Basso-Fossali ed. 2006; Fontanille 2008). This caused the original sociosemiotics to vanish since it gave both elements *a strong ontological value*, thus contravening the *constructivist postulate* of semiotics and above all of sociosemiotics. Texts became artefacts again, something that in our culture is recognisable as meaningful, so much so that they can also be easily defined as what in everyday language is usually called 'texts', that is a book or a similar empirical sample of written communication. On the other hand, a practice, according to this hypothesis, is a social, cognitive and sensorial behaviour such as reading a book, watching a movie, tasting a food, dancing, but also a marketing strategy, a political campaign and so on. It is easy to see how the ghost of the classic *episteme* (Foucault 1970) appears again, insisting in opposing words to things, discourses to facts, representation to reality, fiction to truth, even if semiotics, till then, had fought this kind of episteme, and so had phenomenological philosophy and many researches in human and social sciences of the twentieth century.

When sociosemiotics gradually broadened its field of textual investigation, moving from the work-text to the model-text, it actually brought into question the very use of the word 'text' as the name designating the specific object of investigation of semiotic analysis. This subsequently caused some scholars to go 'over the text' towards other semiotic entities such as individual activities (e.g. cooking, smoking a cigar, walking), ritual activities (e.g. shopping, speaking on the phone), or even more intimate experiences (e.g. dancing, laughing together, tasting a wine). In this new research perspective, there are semiotic objects 'beyond the text' that the theory needs to account for by creating new and more sophisticated descriptive models. For example, the solicitations coming from the new technologies such as the Internet, the *remake*, *remix* and *mash-up* of contemporary cross- and inter-media practices, could not be analysed by traditional semiotics, i.e. textual semiotics, because traditional semiotics was too tied to the text. Here are some paragraphs from an article by Jacques Fontanille (2006b: 1–4, my translation): "The slogan 'Outside the text no salvation' is no longer valid", given that

> semiotic practice itself *has gone far beyond the boundaries of the text* since, about twenty years ago, it began to be interested in architecture, city planning, object design, marketing, but also in the tasting of a wine or of a cigar, seeking a semiotics of situations and of experience arising from different problems of contamination, aesthetic adjustment and hazard.
>
> The moment has arrived, then, for the nature of what semiotics deals with (the object-semiotics) to be redefined in order to acquire in the theory these multiple and necessary researches that had been performed *outside the text*. Those studies can be justified if they are bound to the solidarity between expression and contents, and if they are not, then, an escape from semiosis.

Conclusion:

> If, as Hjelmslev says, for linguistics the data present themselves as a text, the same is not true for semiotics that studies 'objects', 'practices', 'forms of life', creating the structure for whole levels of culture. Hence Greimas' slogan must be rephrased as follows: 'outside the object-semiotics no salvation'. The problem now is how to define the object-semiotics.[1]

---

1 Original version: "Hors du texte point de salut !" est un slogan qui a fait son temps [. . .] la pratique sémiotique elle-même a largement outrepassé les limites textuelles, en s'intéressant, depuis plus de vingt ans, à l'architecture, à l'urbanisme, au design d'objets, aux stratégies de marché ou encore à la dégustation d'un cigare ou d'un vin, et plus généralement à la construction d'une sémiotique des situations et même, aujourd'hui, de l'expérience, à partir d'une problématique de la contagion, de l'ajustement esthésique et de l'aléa. L'heure semble donc venue de redéfinir la nature de ce dont s'occupe la sémiotique (les « sémiotiques-objets »), [. . .] pour assumer théoriquement ces multiples et nécessaires recherches conduites *hors du texte* [c.m.], recherches qui se

Having abandoned the so called 'textualist' position, Fontanille re-ontologises the notion of text: that is, a concept of text as a work-object that needs to be studied in its immanence, grounding on its prior closure. It is thus possible, according to Fontanille (2008), to go 'beyond the text' towards other object-semiotics that need to be defined as such, but which certainly cannot be assimilated to the founding principles of textuality nor to its standard formats. There would be 'an out-of-the-text object' that semiotics could analyse without losing its epistemological 'salvation'. Then, in semiotic studies, the concept of text is in a crisis: it has been questioned in order to let the theory progress, but the theory must address the need for a broader field of investigation and it ends up going back in time and regressing on itself. Sociosemiotics itself is now in the doldrums, the very science that in advanced semio-linguistic studies had to solve the theoretical problems of structuralist semiology: a theory that, speaking in terms of code and sign, depicted the science of signification as an extension of Saussurean linguistic models to non-linguistic social objects – from fashion to advertising, from cinema to the press and so on. For many semioticians, nowadays, the semiotics of text is what art was for Hegel – 'a thing of the past' – because it is not able to account for the new issues of contemporary media culture (where the 'closed' text has no place anymore) nor for the new epistemological needs of semiotic theory (i.e. sociosemiotics, the study of perception, of the body, of new communication technologies, etc.). The awkwardness this situation brings about leads to many questions: does a semiotic analysis of social phenomena lead necessarily beyond the text? What is to be sought? What is there outside the text? Are there contexts? Are there situations? Are there life-forms? Is there experience? Otherwise, on behalf of what concept of text is it necessary to stay within these boundaries? In what situation is it possible to maintain the idea of text as a formal model to describe every object of study of semiotics? Or even better: in what situation is it possible to bring into semiotics another famous slogan, Derrida's (1988) *il n'y a pas de hors-texte* [there is nothing out-of-the-text] – that has been greatly discussed in literary theory and in 'post-structuralist' philosophy?

Sociosemiotics, together with plain semiotics, seems to be at a juncture, awkward in a way, decisive in another. The choice must be well meditated

---

justifient dans la mesure où elles se soumettent à la contrainte minimale d'une solidarité entre expressions et contenus et ne constituent pas des escapades « hors sémiose ». [. . .] s'il est vrai, comme le dit Hjelmslev, que les données du linguiste se présentent comme du « texte », cela n'est plus vrai pour le sémioticien, qui a affaire aussi à des « objets », à des « pratiques » ou à des « formes de vie » qui structurent des pans entiers de la culture. Le slogan greimassien devrait être reformulé aujourd'hui ainsi : « Hors des sémiotiques-objets, point de salut ! », à charge pour nous de définir ce que sont ces « sémiotiques-objets ».

upon. It is now necessary to make a rapid acknowledgment of how the semiotic concept of text was born.

3 From philology to linguistics

Basically, the scientific interest in textuality follows a two-way path: on the one side, there are the philological studies that developed the 'criticism of the text'; on the other side, the linguistic studies that created a real text linguistics alongside a literary criticism inherent to the work. I will now discuss which elements of these two directions are relevant to my focus on the semiotic notion of text.

3.1 A brilliant utopia

Philological research, in general, looks for the authenticity of a text belonging to a distant past that has been lost through historical, social and cultural events. The philological action has a precise goal (to patiently reconstruct an original text of which there is no trace or even no physical presence) and precise aesthetic and metaphysical ideas (the authorial function, unequivocalness of meaning, belief in an existing objectivity and ultimate truth). Philology works with the text as a physical medium – a book, a codex, a manuscript, etc. and, by reading it and confronting it with any other existing copy, it tries to retrace the 'ideal text', the one originally written by an author or, if this is not possible, the closest to the original with the aims of that period, that genre, that style, etc. The link between the empirical text and the ideal Text is, therefore, interrupted and inconsistent and the distance from one to the other cannot be overcome. The accuracy of the philologist, whose aim is to identify the empirical text with the ideal Text, ends up in a brilliant but unfeasible utopia, not just because of a possible inexperience of the researcher, statistical reasons or the historical and hermeneutical diffraction, but basically because of a constitutive principle of language and signification: the same principle stating that the relationship between the two layers of language is not an arbitrary union between any physical medium and a pure theoretical form, but the *mutual presupposition of two forms re-structuring two substances*. A possible perfect reconstruction of the material text (belonging to the field of expressive substance) does not necessarily guarantee the reconstruction of the ideal Text (belonging to the field of forms of expression and, even more, of contents).

It is no coincidence that, although he acknowledged the importance of the philological attitude and experience for semiotic studies, Segre (1979) reasserted the fact that no text (in its empirical written form) can be identified with 'the

Text' (as it was produced by historical and cultural events), even when we casually find the original manuscript by the author. This happens partly because every transcription, including the author's one, is a modification of the original Text, but also because the reading, including that of the textual criticism, is culturally placed, and therefore it is subject to ever-changing cultural codes. There is always a gap between communicative input and output. A text, according to Segre (1979) following Avalle (1973), is not a thing or an object, it is not a physical medium but a 'mental image' that lives alternatively in the author's mind, in the reader's mind and in the philologist's mind, always negotiating between individual versus common, historical versus social elements. The philologists' aim, then, is to establish a new image of the text, making clear what is relevant to their reconstructive analysis (e.g. personal style, genre, historical period, etc.) and taking their own responsibility. And this is true to such an extent that, as Segre says, *philologists do not find given texts that have gone lost, but (re)build them*, fixing themselves the boundaries of texts, their meanings and their cultural values. This leads philology, though for hermeneutical and pragmatic reasons, to an anti-ontological attitude toward the text, against popular opinion. "Every text" Segre (1979: 34, my trans.) writes, "is the voice of a far-away world we are seeking to reconstruct". There is no written piece but a voice, there is no author but a far-away world: in this way the text begins to relinquish its written medium, to acquire a socio-anthropological depth of great significance.

3.2 A negative entity

It is striking how linguistic studies arrived relatively late to the concept of text, linking it with the theoretical results of the practices and meditation of philology. Linguistics becomes aware of the concept of text just when it realises, thanks to Hjelmslev (1961), that the study of languages as formal systems ("an autonomous entity of internal dependencies") goes beyond the opposition between orality and writing. This is a opposition between substances, not forms, and it is therefore unrelated to an analysis of the structure of any language. According to Hjelmslev, the text is the actualisation of the system, it is the starting point for the investigation work, regardless of the substances of expression it may take. Hjelmslev's (1961) famous definition of text – "a syntagmatic whose chains, if expanded indefinitely, are manifested by all purports" – makes it possible to create a link with semiotics as the study of any system of signification, be it linguistic or non-linguistic. To semiotics, substances are not relevant, forms are. This definition, however, does not clarify, strictly speaking, what a text is nor what the essential conditions for its existence are.

Text linguistics takes a step forward on this way in its many theoretical and methodological variations (see van Dijk 1977, De Beaugrande & Dressler 1981; Halliday & Hasan 1985). It takes, as a starting point, the assumption that it is necessary to exceed the boundaries of the sentence that the old normative tradition of grammar, strengthened by the success of Chomsky's generative approach, had set as the only and ultimate object of investigation. What happens beyond the sentence? Which forms and dimensions does a linguistic enunciation take if it does not exhaust its meaning within the sentence dimension? Which formal elements work to put together a group of sentences in order to create one superior linguistic unit, i.e. the text? This brings us to identify a set of rules (anaphora, deixis, co-reference, topicalisation, isotopy, etc.) and principles (coherence, cohesion, informativity, etc.) that, by giving new relevance to some procedures of the ancient rhetoric art, end up facing a double problem. On the one hand, the transphrastic analysis still considers the sentence as the central element of language, thus hypostasising it. This leads to a concept of the text as a coherent sequence of sentences, a kind of 'super-sentence' without any real grammatical or semantic autonomy. On the other hand, since it looks for the text as a stack of sentences, even if a coherent and cohesive one, text linguistics is bound to turn to criteria and procedures working at a higher level than linguistic order itself, such as the communicative intention of the speaker, the semantic acceptability for the receiver, the psychological and cognitive nature of the participants, the situational and pragmatic context, the hermeneutic condition of interpretation, the referential background, and so on. The text hangs between the inside and the outside of language, as a kind of ambiguous interface between saying and doing, words and things, and it actually has no place on either side.

It is easy to see that, in both cases, the text is appealed to as a necessity, but it is not actually identified and limited as an autonomous object of analysis for linguistics. Both Hjelmslev and text linguistics define it always in the negative (either to contrast the primacy of writing, or to go beyond the boundaries of the sentence), but they never recognize the role it actually plays in linguistic communication: that of a 'natural' entity, a full enunciation, a global linguistic product that gives form and meaning to local elements inside itself (including sentences), but that also gives consistency and value to entities outside itself, such as the discursive genre, the situational context, the hermeneutical horizon. For the text to take this central position between its intrinsic elements and its external polarities, it must be dealt with mainly in its semantic dimension, supported by specific forms of expression. That is to say that the text has to be seen as a 'unit of signification', a global communicative unit with its own internal structure, a specific hierarchy of parts, its own cognitive and pragmatic dimension and without dependent relationships with any external reality.

4 Aesthetics and methodology

Many researchers in the linguistic field have tried, and are still trying, to confront the problem from a semantic perspective, either by proposing new theoretical hypotheses (discourse analysis, theory of speech acts, narratology) or by resuming ideas of more traditional disciplines such as stylistics, rhetoric, poetics, literary criticism. Mature semiotics wanted to find a solution for the theoretical deadlocks caused by the concepts of sign and code and, above all, it wanted to be independent of the linguistic model. For this purpose, it tried to make these different research pathways communicate with each other by putting the text in the centre as its specific object of study. In other words, it moved from semiotics of codes to semiotics of texts. The text, in semiotics, becomes *the overall configuration of meaning that, by means of some expressive medium, guarantees the generation, circulation and interpretation of cultural and social meanings*. Every culture institutionalises such a configuration in its own way and semiotic analysis must be able to understand and describe it.

In semiotic research, however, the passage from the study of signs to the study of texts, though producing good results, has not been explained clearly from the theoretical and methodological point of view. Between the late 70s and the early 80s, the term *code* gradually gave way to the term *text*, but nobody clarified the reason for this terminological transition expressing an underlying conceptual modification, or even a completely new epistemological predisposition. Semioticians, making explicit or implicit reference to different research pathways – from philology to sociology of media, from glossematics to text linguistics, from pragmatics of communication to discourse analysis, from the theory of speech acts to rhetoric and hermeneutics – have used the term *text* often without defining it, often referring to entities and phenomena that could not be easily compared: on one *occasion text* was with reference to the linguistic model, on another with reference to the communicational model, the literary model and so on. Hence the many misunderstandings mentioned above (for a reconstruction of this passage from code to text, see Manetti and Violi 1979; Eco 1984).

4.1 From Work to Text

In order to resume my reconstruction and to explain some still unclear passages, let me now take into consideration two key points. The first one is the opposition *Work* vs *Text* that Roland Barthes (1986) claimed. Such a distinction, though full of references from Lacanian psychoanalysis to *Tel Quel* artistic experimentalism as well as to the political and philosophical debate of the 70s, is extremely clear and useful now, since it allows us to understand, among other things, what textuality meant in that period. This is possible because, in the first place, it is a relational and systemic definition that does not define the text in itself, but according to its relationship with its symmetric contrary (the work); and, secondly, because it is not a distinction between things, but between views: because it does not take into consideration ontological assets, but theoretical and methodological ones. Work and text start from two divergent points of view and reach for the same object, ending up by building it in deeply different ways. The *Work* is the proper literary work following all the rules of the aesthetic aura: authoriality, sacredness, unequivocalness of meaning, insertion into a pre-set framework of genres, references to value and anthropological criteria, a certain presumption of objectivity. The *Text* is a different vision of the same object that mainly aims at eliminating this alleged objectivity, thus removing also the possibility of putting it on a shelf, of identifying it as belonging to a genre, connected with an author, limited within an aesthetic view. The text, according to Barthes (1975b), is the acknowledgment of the dominance of method over object, a 'methodological field' that inevitably multiplies possible views, interpretations and meanings. That is why, as Barthes used to say, *the text is plural*: not so much because it is an open work subject to multiple interpretations, rather because it admits, with strategic modesty, other interpretations and methods as equally possible, providing their conditions of exercise. They are methods not things, views not empirical phenomena. If the work is consumed, the text can be productive. Hence the idea that a text is not opposed to practices, but is itself a 'practice of signification', a production (Kristeva 1980), a way to let streams of meaning flow and, in doing so, takes on itself the task of explaining their formal procedures, their linguistic turns, their rhetorical games. The text has no pre-set boundaries, no objective limits, because it has in itself multiple 'perspectives of quotes' (Barthes 1975b), countless other texts thriving in mutual exchange with it, inside and outside. The text fades into intertextuality, as intertextuality itself is already within the text. The cultural structure, more or less institutionalised, more or less quiet, sets the boundaries of the text and tightens outside it the chains of intertextuality. The methodological view of the researcher must identify what this structure is and, if need be, what its political values are. The immanent view of the critic that has been a focal point from the Russian formalists onward, does not mean the blind acceptance of the boundaries set to the work by cultural institutions; it means, instead, the problematic construction of liminal thresholds, the responsibility of which is on analysts themselves. *Immanency is created, not given*. This is the framework into which Barthes's textual analytic practice must be inscribed. In works such as *S/Z* (Barthes 1975b) he claims the 'difference' of every single text – a tale of Balzac's or a Japanese *haiku* (Barthes 1982a) – compared to the deep structures it has in common with other similar texts. This is not a new aesthetic dress for the old concept of work, but a theoretical option that emphasizes the methodological procedures whose aim is to bring to the surface the multiple – rhetoric, stylistic, linguistic, etc. – forms whence the stream of meaning originates.

4.2 Discourse and narrativity

The other key point we have to look upon, seemingly far from Barthes's position, is the study of discourse and narrativity. Like the previous one, such a study perspective originates in Russian formalism and Propp's works (Todorov ed. 1966; Propp 1958) and it arrives at the construction of a narrative grammar and a discursive syntax (Greimas 1987c), after crossing comparative studies in ethnology, history of religions, folklore, literature and paraliterature, mass communication, cinema, history of art, and so on. What is interesting for us now, however, is the moment when the study of discourse and the study of narratives enters into the research field of semiotics to face the semantic problems of textuality. Before transphrastic syntax failure, *the only way to identify text features seemed to be the reconstruction of its semantic structure* in a way that cannot be a mere list of co-referential and anaphoric links between words and sentences, since the text has a uniform articulation of meaning of its own, referring to different possible typologies (description, argumentation, demonstration, etc.), and where narrativity plays a seemingly fundamental role. The reason for this is not just that texts have many explicit narrative values, but that even seemingly non-narrative texts follow forms and procedures typical of narratives: internal processuality, controversies among the forces involved, final transformation and so on. So Eco (1979) is compelled to say that, though it is demonstrated in geometrical order, even Spinoza's *Ethica* has an underlying narrative structure; and in the same way, Greimas (1988) notices that a particular description by Maupassant is actually a specific move within a narrative. Both call attention to this element because they want to validate the idea of an autonomous textuality that would be free from the linguistic surface, but tied to the semiotic procedures structuring its deepest meaning.

Similarly, underneath the text there is a discourse involved in the construction of the uniform semantic structure of the text. The terms *text* and *discourse* are often thought of as synonyms since both express linguistic occurrences greater than the sentence. Sometimes, according to different research pathways, scholars preferred the one and sometimes the other, sometimes they even ended up attributing the written substance to the first and the spoken substance to the second. In many other cases the term *discourse* is used to call attention to the communicative dimension of the linguistic text by focusing on how linguistic forms are used in communicative situations (pragmatics), or to how different languages put into grammar their situations of emission and reception, thus pre-setting communication scenarios (the theory of enunciation). In both ways, the general trend is to define a set of specific rules of a purely semantic nature that constitute textuality and make it the basic linguistic unit, the starting point and the goal of any communicative phenomenon.

Every text, then, chooses a semantic structure transcending itself, thus making it similar to other texts and differentiating it at the same time. The text is, on the one hand, *the empirical evidence of a deep narrative or discourse, the tip of the iceberg of a wider culture* of which it is a witness. This, however, does not mean that the text is nothing but an application of those deep models that include it and make it similar to any other text, erasing any particular feature; instead, it keeps its own specificity, thanks to the unique way in which it locally makes explicit larger cultural structures. Therefore, the existence of discursive models underlying the text guarantees its circulation within the cultural sphere, but it also guarantees the autonomy of meaning of the text and its relationships with other texts.

As we can see, there already seem to be many elements that will allow the passage of the concept of text from the linguistic and literary field to the wider semantic and semiotic, social and cultural field: independence from the written, and more generally linguistic model; identification of the text as the main object of the semio-linguistic investigation; dominance of the method and consequent 'de-ontologisation' of textuality; multiplication of the layers of meaning; primacy of semantics and so on. This passage allows the concept to be used as a formal model to study not only communicative non-verbal occurrences (images, movies, songs, different kinds of audio-visual material, etc . . .), as is natural of any semiotic science, but also semiotic phenomena such as advertising campaigns, political strategies, everyday rituals, wide spaces and socialisation forms within them, urban centres, practices of use of media and of goods and services, sensorial and corporeal experiences and so on. These phenomena would not be analysed *as if* they were texts – imagining a metaphorical extension of the linguistic text to the non-linguistic one – but *because they are* texts. This would be done by reconstructing their implicit textuality – their underlying semantic, discursive and narrative structure – the same textuality that gives them a meaning, makes this meaning spread in social culture, appearing and disappearing, always changing, giving origin to understandings and misunderstandings, translations and infidelities.

At the beginning, however, nobody took this step completely, or at least nobody did it explicitly. A movement in this direction can be discovered in applied studies, in field research, in the many analyses that have been proposed in different fields and languages (see Metz 1974, 2011, 2016; Bettetini 1984; Calabrese 1985, 1999; Marin 1995, 2002). Nevertheless, this step has never been discussed in theoretical writings nor in essays, never before sociosemiotics came along, and maybe not even then.

5 Hermeneutics, deconstructionism, textualism

While philology and linguistics seem to insist on the material side of the text, hermeneutics, in its many forms, works on the opposite side: the one of sense and meaning. It seems as if each of the two disciplinary fields took on itself one half of the text – 'writing' the first and 'spirit' the second – thus replicating for language the ancient dualism that exists in respect of the subjectivity between the body and the mind. Consequently, according to hermeneutics, contents of texts do not originate from a constitutive solidarity with the plane of expression, as Saussure and Hjelmslev said, but they come out of a wide series of different procedures (psychological, logical, cognitive, affective, etc.) which are mainly autonomous from their constitutive linguistic base and are not bound to any kind of semiosis. So those philosophical studies so greatly discussed – from Heidegger's and Gadamer's to Ricoeur's, via Derrida's and Rorty's studies (to mention just the main reference points in the twentieth century debate) – generate a sort of paradox: they claim the essential importance of language not just to theoretical thinking, but also as a necessary mediation for any form of human and social experience. At the same time, however, the above-mentioned studies seem to take into no consideration semio-linguistic studies that try to describe the internal functioning of language. On the one hand, they insist on the idea of philosophical writing; on the other, they do not accept the formal description of such a writing. On the one hand, they maintain that there is nothing outside the text; on the other, they cut off the very possibility of investigating what is inside the text, what it is made up for, what its articulations are, what are its reasons. In other words, even if they confirm Heidegger's idea of language as 'the House of Being', not a single scholar seems to be interested in the architecture of this House, in the foundation, in how the rooms are divided or what furniture is used . . .

This paradox can be explained by the refusal of scientificity and objectivism that semiotic and linguistic disciplines still claim as their peculiar characteristic. It is well known that hermeneutics does not refuse scientificity, like Descartes' rationalist methodology (or, as they usually call it, 'epistemology'), whereas many forms of structuralism keep it or, at least, do not refuse it explicitly. Hence comes the rupture between hermeneutics and linguistics while, conversely, they could and should meet in the name of the text, thus enriching each other: the first by bringing with it the interest in textual structures; the second by explicating its epistemological position. The result would be a 'material hermeneutics' (Schleiermacher, in Rastier 2001): that is, a study perspective focusing on the problem of textuality in its many aspects (the material-philological one, the formal-linguistic one, the socio-cultural one, the cognitive-interpretative one, etc.) and, at the same time, exceeding the naive narrowness of empiricism and the indeterminate digressions of structuralism. Let us have a deeper look, then, at what hermeneutic studies bring to the semiotic of textuality and how.

5.1 Contributions of hermeneutics to sociosemiotics

Hermeneutics is, as everybody knows, a discipline strongly dependent on the *presence* of existing texts. At the beginning they were sacred and legal texts; later on they became literary and philosophical texts; but they constitute, in any case, a *cultural tradition* that settles down, becomes *institutional* and *stable* within texts. From this perspective, the text is a written manifestation, a strengthened *trace*, passed on through generations, of a cultural choice which transcends the text. This brings about theoretical choices such as those concerning the 'hermeneutic circle' (refusal of any gnoseological *tabula rasa* in the name of an unavoidable pre-cognition of the world) and the 'linguistic characterisation of being' or rather of its understanding as the unavoidable mediation of words for any access to the world. So pre-cognition is linguistic because it refers to texts that tradition has passed on and that constitute tradition itself since they use what, by definition, settles memory (or, we may say, makes oblivion useless): writing.

Nothing seems farther from the sociosemiotic perspective. While hermeneutics claims the supremacy of language and of written texts – that, by definition, are more authoritative than others – according to sociosemiotics any matter of the world, in a given cultural situation, could carry with it human and social contents. In other words, from a semiotic point of view, every culture speaks of itself to itself, constantly increasing the number of languages it needs, and those languages intertwine with each other according to ever-changing values in space and time. On a closer look, however, the hermeneutic view can offer sociosemiotics different theoretical contributions and epistemological clarifications. The main contribution is, as I said before, the refusal of any pre-set scientificity deriving from a naive ontologism and from a representative and referential concept of language: *'facts' never come first, because they are the result of a cultural construction* that necessarily involves the beholder; every scientific stance is placed within a well-defined historical and cultural situation participating in the creation of its object of knowledge. From the semiotic point of view, that leads to the idea that the text is not an empirical reality given once and for all but a cultural structure changing its meaning and form depending on the historical and social conditions in which it was born, and on the discursive environment in which it spreads. That also leads to questioning the descriptive meta-language because, since it presents itself as a text that has to be built and analysed from the semiotic perspective, it ceases to be a pure and abstract model useful to performing whatever kind of analysis. So-called 'textualism', according to Barthes and Derrida, claims that there is no actual difference in the nature of the two things – they just have a different discursive position that is, as such, vulnerable to repeated variations.

Another important contribution of hermeneutics to the constitution of rigorous sociosemiotics, somewhat implicit in what I said before, is the constant presence of tradition as a background for every textual occurrence and, at the same time, as a goal for every act of interpretive understanding. This tradition, however, is not to be considered as a given and undisputed model, more or less sacred and authoritative, as often happens in hermeneutic studies. Such studies, even those deriving from deconstructionist positions, end up hypostasizing poetic and literary textuality as the 'creative' solution to any authoritarian position. Tradition has to be replaced, instead, with a wider and more generic 'semiosphere', wherein texts find their place in a constant dialogue with each other, being at once the discourse and at once the object of the discourse. There are texts and meta-texts that, thanks to the mediation of any expression-substance, gain different values and functions according to their anthropological horizons. Even if culture is the reservoir and manufacturer of texts, the texts within it are not equal, nor do they have the same instrumental and discursive value.

5.2 The outside-text

Let us take into consideration two controversial points about the relationship between semiotics and hermeneutics. The first one is the famous statement by Jacques Derrida (1988) that has become the shallow slogan of deconstructionism and, as such, has given origin to any kind of rumours, claiming '*il n'y a pas de hors-texte*'. As I mentioned before, from a sociosemiotic point of view this statement is perfectly sharable. Sure enough, it must not be considered as recalling the hermeneutics of tradition that sees Texts as the beginning and the end of any horizon of thought; nor does it have to be regarded just as a claim on a pre-set closure of the linguistic universe that would lead to the refusal of any referent outside the language system and, consequently, of any contribution by psychological, social, economic and other situational 'contexts'. Such a statement is not an amplification of structuralism, that considers language, as Hjelmslev does, to be an autonomous entity with a peculiar internal structure (opposed to the idea of a correspondence between words and things as the ultimate criterion to find truth); and it is not to be considered just as a confirmation of the research program of early Russian Formalism or of American New Criticism, both committed in the study of literary works regardless of production and reception contexts. And, above all, 'Il n'y a pas de hors-texte' does not have to be translated as 'there is nothing outside the text'. This is how Gayatri Spivak translated Derrida's statement, causing a deconstructionist interpretation to spread. An interpretation that, as has been noted, presents every characteristic of an instinctive metaphysics. As I will demonstrate, the statement needs to be interpreted in this way, using the hyphenated phrase 'outside-text': *there is no outside-text, because as one comes out of a text one enters another, and then another, and one more after that and so on forever.* There is no human signification of any nature that would not take the form of a text, that would not belong to the textual order, regardless of the peculiar expression-substance this textual form chooses to circulate through. Every context is a co-text, or it is not relevant.

To deeply understand this statement, just take a closer look at the (textual) place in which it has been expressed: an intense discussion with Jean-Jacques Rousseau. In *De la Grammatologie* [*Of Grammatology*], Derrida (1988) retraces the thought of this author and, in doing so, he makes a genuinely semiotic operation. In fact, he brings into question the complete works of Rousseau (defined as 'the epoch of Rousseau' so that in 'epoch' resounds *'epoché'*) including both philosophical and narrative texts. Moreover, within these works, Derrida considers, beside the 'spirit' of the texts (the 'thought' of Rousseau), their 'writing': that is their linguistic-expressive form, but also their argumentative and narrative one that, by conveying this thought, gives it a peculiar semantic conformation. The result is an analysis (or a deconstruction) of the corpus of Rousseau's texts which – here comes the point – integrates the reconstruction of his complex thought. It is an analysis that, if and when it needs to leave every single text, always finds another text to support it and to bring it to a better understanding.

Here follows the fragment from Derrida's work where we find some important considerations about 'questions of method' inherent to how Derrida (1988: 159) reads Rousseau:

> If reading must not be content with doubling the text, it cannot legitimately transgress the text toward something other than it, toward a referent (a reality that is metaphysical, historical, psychobiographical, etc.) or toward a signified outside the text whose content could take place, could have taken place outside the language, that is to say, in the sense that we give here to that word, outside of writing in general. . . . *There is nothing outside of the text* [there is no outside-text; *il n'y a pas de hors-texte*]. And that is neither because

Jean-Jacques' life, or the existence of Mamma and Thérèse *themselves*, is not of prime interest to us, nor because we have access to their so-called "real" existence only in the text and we have neither any means of altering this, nor any right to neglect this limitation. All reasons of this type would already be sufficient, to be sure, but there are more radical reasons. What we have tried to show by following the guiding lines of the "dangerous supplement", is that in what one calls the real life of this existences "of flesh and bone", beyond and behind what one believes can be circumscribed as Rousseau's text, there has never been anything but writing; there have never been anything but supplements, substitutive significations which could only come forth in a chain of differential references, the "real" supervening, and being added only while taking on meaning from a trace and from an invocation of the supplement, etc. And thus to infinity, for we have read, *in the text*, that the absolute present, Nature, that which words like "real mother" name, have always already escaped, have never existed; that what opens meaning and language is writing as the disappearance of natural presence.[2]

As everybody can see in the previous citation, *'il n'y a pas the hors texte"* [there is no outside-text'] *does not mean that 'there is nothing outside the text'*, but, on the very contrary, *it means that there is no human or social experiences of anything other than a textual kind*. What we call the world of real life, actual living, affectivity, corporeal and carnal intimate subjectivity – which phenomenological hermeneutics highlights – can be understood – that is to say, can be experienced meaningfully – only if it is screened by a textual grid (that, according to Derrida, has a 'written' nature). Beyond any philosophical value of deconstruc-

---

2 Original version: "« Et pourtant, si la lecture ne doit pas se contenter de dédoubler le texte, elle ne peut légitimement transgresser le texte vers autre chose que lui, vers un référent (réalité métaphysique, historique, psycho-biographique, etc.) ou vers un signifié hors texte dont le contenu pourrait avoir lieu, airait pur avoir lieu hors de la langue, c'est-à-dire, au sens que nous donnons ici à ce mot, hors de l'écriture en général. C'est pourquoi les considérations méthodologiques que nous risquons ici sur un exemple sont étroitement dépendantes des proposition générales que nous avons élabores plus haut, quant à l'absence du référent ou du signifié transcendantal. *Il n'y a pas de hors-texte*. Er cela non parce que la vie de Jean-Jacques ne nous intéresse pas d'abord, ni l'existence de Maman ou de Thérèse *elles-mêmes*, ni parce que nous n'avons accès à leur existence dire 'réelle' que dans le texte et que nous n'avons aucun moyen de faire autrement, ni aucun droit de négliger cette limitation. Ce que nous avons tenté de démontrer en suivant le fil conducteur de 'supplément dangereux', c'est que dans ce qu'on appelle la vie réelle de ces existences 'en chair et en os', au-delà de ce qu'on croit pouvoir circonscrire comme l'œuvre de Rousseau et, derrière elle, il n'y a jamais eu de l'écriture ; il n'y a jamais que des suppléments, des significations différentielles, le 'réel' ne survenant, ne s'ajoutant qu'en prenant sens à partir d'une trace et d'un appel de supplément, etc. Et ainsi à l'infini car nous avons lu, *dans le texte*, que le présent absolu, la nature, ce que nomment les mots de 'mère réelle', etc., se sont déjà dérobés, n'ont jamais existé ; que ce qui ouvre le sens et le langage, c'est cette écriture comme disparition de la présence naturelle »" (Derrida 1967 : 227–228).

tionist grammatology, what is central is the idea of a wide-ranging textuality, wherein it is impossible to tell the text apart from its exterior comment. What Derrida, following certain of Rousseau's verbal tics, calls 'dangerous supplement' is something already present in the writing and its textual revelation, something the critical reading must not intuitively reveal but methodologically 'produce'. By a tactical reading-between-the-lines of *Of Grammatology*, I think we are facing mainly the *enunciative dimension of the text*, a dimension that is both inside and outside the text itself allowing not just the possibility to identify, but also to construct discursive subjectivity. The *I* of Rousseau, emerging either from autobiographical novels or from philosophical essays, is 'inscribed in a determined textual system'; that is to say, it is produced by unequal enunciation structures inside different texts. This *inside*, however, as Derrida insists on saying, cannot be discussed apart from the *outside*, both because every text is part of a textual chain that by confirming the inside transcends it, and because it is at the borders of the text, here in the sense of enunciation (that is narrative, reflexive, etc.), that its actual enunciation takes place, that is the process of putting it into discourse. Such a process was already present in the form of grammar rules within language and the process also allows the (dynamic and changing) creation of intertextual nets. I think we can read the following strong methodological statement as leading to discursivity, even if it is never named explicitly:

> When we speak of the writer and of the encompassing power of the language to which he is subject, we are not only thinking of the writer in literature. The philosopher, the chronicler, the theoretician in general, and at the limit everyone writing, is thus taken by surprise. But in each case, the person writing is inscribed in determined textual system. Even if there is never a pure signified, there are different relationships as to that which, from the signifier, is presented as the irreducible stratum of the signified. For example, the philosophical text, although it is in fact always written, includes, precisely as its philosophical specificity, the project of effacing itself in the face of the signified content which it transports and in general teaches. Reading should be aware of this project, even if, in the last analysis, it intends to expose the project's failure. The entire history of text, and within it the history of literary forms in the West, should be studied from this point of view.[3]　　(Derrida 1967: 160–161)

That is the reason why Rousseau, or anyone else, "cannot be separated from the system of his own writing". That is also the reason why the writer 'inscribed

---

[3] Original version : "Quand nous parlons de l'écrivain et du surplomb de la langue auquel il est soumis, nous ne pensons pas seulement à l'écrivain dans la littérature. Le philosophe, le chroniqueur, le théoricien en général, et à la limite tour écrivant est ainsi surpris. Mais, dans chaque cas, l'écrivant est inscrit dans un système textuel déterminé. Même s'il n'y jamais de signifié pur, il y a ses rapports différents quant à ce qui du signifiant *se donne* comme

textuality in the text' and why 'wherever we are' we are 'in a text where we already believe ourselves to be'. In other words, that is the reason why there is no outside-text. Later on, textualism, as Richard Rorty (1979) defined it (considering it as the 'ironical heir' of the Hegelianism of the 19th century), interpreted in different ways Derrida's ideas on this point, giving them different values and interpretations that, nevertheless, followed one trend towards the hypostasis of textuality as the written dimension (not just in literary works, but also for philosophical, historical, social sciences). Such hypostasis led to the claim of the total interpretive freedom of the reader, free from any metalinguistic binding, that originates from the idea – opposite to the one proposed here – according to which no real meaning of the text exists. What exists are, in a hermeneutic way, its many effects in the theatre of history. In conclusion, textualism, particularly in its *strong* version, as Rorty himself defines it, has no respect for the text as such, nor is it interested in the text's procedures or internal functioning. Thus, the text dons an ambiguous guise, opposing it drastically both to any textual criticism (in the traditional sense of philology) and to any text semiotics (that, because of the ignorance of history, will be termed 'textualist semiotics'). Yet the criticism of the concept of context as extra-linguistic exteriority and guarantee of truth of the ultimate meaning of the text, does not imply giving up any textual signification. On the contrary, it strengthens the textual signification, or even doubles it by identifying this 'internal supplement' within the text that is not a funny oxymoron to play with but, I think, the discursivity of the text itself, its ability to build forms of subjectivity and processes of inter-subjectivity on the plane of enunciation.

5.3 The hermeneutic arch

Paul Ricoeur's analysis follows a completely different path, explicitly going against the textualist deconstructionism in the name of Derrida, with its many followers. What is interesting as we trace back the origin of the concept of text (and its analytical and interpretive practice), is that Ricoeur's hermeneutics brings closer together the subjective *understanding* of the sciences of spirit and the objective *explanation* of the sciences of nature (using Dilthey's classical terms), placing

---

irréductible de signifié. Par exemple, le texte philosophique, bien qu'il soit en fait toujours écrit, comporte, précisément comme sa spécificité philosophique, le projet de s'effacer devant le contenu signifié qu'il transporte et en général enseigne. La lecture doit tenir compte de ce propos, même si, en dernière analyse, elle étend faire apparaître son échec. Or toute l'histoire des textes, et en elle l'histoire de formes littéraires en Occident, doit être étudiée de ce point de vue" (Derrida 1967 : 229)

them not as opposing but as complementary concepts. Explanation expands understanding and vice versa. Such complementarity is actually justified by the concept of text and its undeniable centrality. The reasons for this are, in the first place that, according to Ricoeur (1974), language as such cannot be considered like any other natural entity and, consequently, linguistics cannot be limited to a mere naturalistic explanation; in the second place, that the text, as the written deferment of the communicative-linguistic process, establishes a kind of hermeneutic game where explanation and understanding complement each other. According to Ricoeur (1974), writing is not a mere transfer of oral speech – preceding it by principles – into a game of signs that, at most, strengthens its memory. The text (always written, as in Derrida), actually delays the dialogic property of the word, neutralizing the game of reciprocal replies in which the meaning springs directly from the life of the participants. Thus, discourse loses any reference both to the situational context and the natural world; yet it does not efface referential reality completely. Every text creates its own referentiality, a sort of textual world suspended between the empirical reality of the text and the tangible reality of the natural world. Sometimes the text is linked to other similar texts and together they produce a transcendent dimension that Ricoeur (1974) calls – with a still traditional name – literature, literary imagination (we can re-define it in terms of intertextuality). From this point of view, says Ricoeur in agreement with textualism, every text is closed in itself and does not worry about the exterior, the outside-text. On the other hand, in a very specific moment that is but another part of the textual dimension, the written text opens itself again to the world, to life, to subjective experience, to the systems of existential and cultural values outside it: it is the moment of reading. Reading restores a kind of dialogue between the participants of the discourse and thus allows concrete reality to reappear, after the text had put it aside or even deleted it through the act of writing. Reading, according to Ricoeur, is like performing a music score: it arouses the truth of the text thanks to one's own subjectivity, to one's own inner world, to one's own experiences, to one's own values. In this way it restores the continuity with a tradition of textual nature whose meaning, otherwise, would have got lost in the distances of time.

For Ricoeur, the closing and reopening movement of the text is repeated when interpretation is performed, first as a structural analysis of the textual world, closed in itself and suspended above the natural world and after, as the actual interpretation that re-establishes the link between the internal functioning of the text and the external cultural reality, thus recreating a continuity of problems and values, existential experiences and collective lives, and tying a new discourse to that of the text. Hence comes the example of structural analysis of the myth performed by Lévi-Strauss (1963) (that, as we will see later, is not very well conceived from the point of view of textual configuration). Ethnology, says Ricoeur, shows that the myth of Oedipus is a system that becomes aware of human contradictions (kinship, distance between man and beasts, etc.) and tries to balance them at another level: that of mythological narrative. Identifying this internal system of the myth, though, cannot be an end in itself, because it reveals some basic human sets of problems that are as universal as they are unsettling, such as the problem related to the deep relationship between life and death. The Oedipus myth represents the anxiety of origin: the conjoint action of structural analysis and interpretation – defined as 'hermeneutic arch' – reveals that anxiety clearly. The main role, however, is still played by interpretation, an interpretation that is free from psychology and quite objectified, since it is tied in the background to linguistics, that is, to the text of which it wants to reveal the proper intentionality, apparently coinciding with the deep semantic structure of the text itself. The task of structural analysis, which is to retrace the *meaning* of the text, is now complementary to the task of interpretation, which is to identify the text's *directionality*. In both ways, says Ricoeur, the key matter is the *sense* to be re-activated, thus guaranteeing the fusion between the textual and experiential world. In this way the sociosemiotic choice seems more reachable.

5.4 Interpretation and configuration

This can be demonstrated by another of Ricoeur's important discussions, the one about the meaningful action considered as a text, that is to say the possibility of looking at the phenomena studied by human and social sciences as if they were texts because they have the same formal and procedural characteristics – which is also my hypothesis in this book. Ricoeur's (1973) procedure consists of finding a specific hermeneutic object of knowledge and a specific hermeneutic methodology for human sciences. Human sciences look at their objects of knowledge with an attitude that can be compared to the one the interpreter takes while reading a text (according to the dialectics between explanation and understanding). The reason for this is simple: texts and socio-cultural phenomena have the same characteristics. While many human and social sciences aim at an ideal scientificity inspired by the epistemology of natural sciences (hence the methodology of quantities, numbers, statistics, etc.), Ricoeur advocates for the qualitative methodological aspects of those disciplines, seeking to base such qualitative method on a previous analysis of what could be considered as their specific object of study that is created, not given, based on a cultural apperception rather than on empirical knowledge. This is what makes this hermeneutic discussion so important to us, since it is tied deeply to the very sense of sociosemiotics, namely the relationship with the general semiotic theory on one hand, and with human and social sciences on the other. As we will see, *sociosemiotics* does not mean merely extending the semiotic field to involve sociality, nor it is a naive offer to sociology of a professed rigorous methodology. It is, instead, *the critical reconstruction – even in the Kantian sense – of the conditions of possibility of society as an object of scientific knowledge*, it is the discovery of semiotic procedures that make human and social phenomena interesting and relevant to social analysis; it is the elaboration of semiotic reasons for something to be thought of as a social object (see Landowski 1989; Marrone 2001). Consequently, Ricoeur's attempt to put social phenomena and texts together us very interesting.

Let us follow his argument. As the written text differs from spoken speech in some basic characteristics (embedding and fixation of the enunciated content, non-ostensive reference, possibility of universal readings), the social events, once becoming an object of sociological investigation, differ from the ordinary actions of individuals in the same basic characteristics: objectification of action in a specific configuration given to interpretation, based on one's own internal connections; detachment of action from agent; consequent autonomy from any subjective responsibility; overtaking of the conditions of production of the action and repositioning of the latter on a socio-cultural and historical level; possibility of a non-contemporary, almost universal interpretation of action. Once these textual characteristics of meaningful action have been fixed, it is possible to handle the object as a text and to identify in its interpretation a dialectic relationship between explanation and understanding. On the one hand the original, naive understanding of meaningful action is expanded by intrinsic, structural explanation of significant social phenomena; on the other hand, such an explanation leads to a deeper understanding of the event, thus connecting the interpreter's world with the significant event world. The exact same thing happens with texts.

What is more valuable about Ricoeur's attempt, though, is the set of preliminary remarks rather than the actual results, tied to a hermeneutics that, even though linguistically founded, still depends on tradition and its new philosophical value. What Ricoeur actually considers social facts are 'meaningful actions' in the Weberian sense of the term: meaningfully-oriented behaviours, where 'meaningful' means 'significant' and provided with a great value in society (and consequently institutionalised), a social importance that goes beyond the peculiarity of individual actions. The comparison with the text is based on such hypostasis and this also influences the very concept of text Ricoeur has in mind: to him, texts are written texts with a certain intrinsic authority, with a cultural value of their own, that cannot be compared at all to the objects of the sociosemiotic analysis, that, on the contrary, acquire no *a priori* value.

What is closer to the theoretical perspective of semiotics more than this discussion about a hermeneutical methodology of social sciences – although it explicitly mentions the term *text* – is Ricoeur's consideration of the link between time and narrative. In his three famous books on the topic, Ricoeur (1984–1988) shows clearly how before the actual narrative (novel, epic poem, tragedy, fairy tale, etc.), at the level of the so-called world of life, a sort of instance of understanding of a narrative nature can be found. For daily life to make sense, to be delimited and understood, it is necessary to organise both actions and events happening to us according to narrative logics, so that some categories that are typical of narratives (intentionality, motivation, value, action and reaction, temporality, etc.) can grant us to act in a meaningful way, to act according to existential projects that prove coherent because they are filtered through a narrative sieve. Narrativity is one of the main dimensions in which texts are built and, according to Ricoeur, it is also the form of our experience. It is not surprising, then, that Ricoeur (1984–1988) is interested in Greimas's semiotics of narrative and text, that, as we will see, seems to provide an answer for many questions the hermeneutics of text leaves unanswered.

6 Semiotics of text

I already mentioned that, in semiotics, the concept of text has been imported from the closest disciplinary fields here identified, in order to try to solve the many difficulties of the first semiology of sign and code. Barthes (1967, 1983) had tried to use some key categories of structural linguistics (langue/parole, signifier/signified, syntagm/paradigm, denotation/connotation) as tools to describe different social objects and phenomena, such as cuisine or furniture, city planning, medicine, petty bourgeois mythologies of mass culture and, above all, fashion. Following these suggestions for research, many scholars tried in a similar way to re-structure different fields of study (painting, photography, theatre, cinema, architecture, design, gestural communication, publicity, etc.) according to the linguistic model, thus distorting the nature of such objects almost to absurdity. Such a trend brings about, for example, terms modelled on 'phonème' such as 'mythème', 'cinème' 'gustème' and so on; or it leads to investigating the possibility of comparing a picture to a linguistic enunciation, or to exploring how many articulations exist in the cinema, what are the possible analogies between words and photograms, or to thinking of the possibility of creating a dictionary of gestures and other theoretical hypotheses that everybody sees now as vaguely amusing cul-de-sacs. This trend ended up by reducing any proper specificity of different semiotic repertoires to the only repertoire unanimously considered as prime – verbal language. The problem, actually, did not lie exclusively in the theoretical hypostasis of the verbal language but, more deeply, it depended on a shared, though unstated, presupposition: that in human cultures there exist different semioses whose difference depends just on the expression matters they use, on the sense organs needed to produce them, on the media needed to transmit them.

The concept of textuality has been given a prominent position in semiotic discussions in order to prevent the same kind of theoretical and methodological difficulties, always keeping an eye on what is happening in text linguistics, narratology, speech acts pragmatics, theory of communication, hermeneutics and deconstructionism (Marrone 2001). The text offers to semioticians the possibility of identifying, as their own object of investigation, communicative units and wide configurations of sense that can be *independent from the different expression-substances* (that, after Hjelmslev, have been considered irrelevant to linguistic analysis), and can also *transcend the strict opposition between pre-set semiotic systems and their single realisation*, between languages and speech acts, between codes and messages, *langue* and *parole*. The text employs given systems but, by using them, it modifies them, it makes them relevant to the communicative environment in which it circulates while absorbing given cultural configurations that allow the construction of its complete and complex structure of sense. The following period, the euphoric decade of the 1970s, saw within the domain of semiotic studies many theoretical experiences and experiments of analysis, one after the other. They were based on the hypothesis according to which textuality has its own rules that are neither exclusively tied to linguistic grammars nor bound to aesthetic idiolects or the subjective readings of the public. On the one hand, the main role was played by literary analysis: in 1970 Barthes studies Balzac's *Sarrasine* showing the multiple pathways of meaning within it (*S/Z*); in 1976 Greimas performed an analogous analysis on Maupassant's *Deux Amis*, thus implicitly establishing the procedures for a semiotic analysis of texts (*Maupassant*); in 1979 Eco proposed his theory of interpretive cooperation of narrative texts explaining it through the example – next to the omnipresent Nerval's *Sylvie* – of Alphonse Allais's *Un Drame Bien Parisien* (*Lector in Fabula*) (see Marrone 2014). On the other hand, at the same time and afterwards, there appeared an increasing number of textual analyses on paintings, movies, theatre pieces, comics, advertisements, newspaper articles, music works, design objects, architectonic works. All of them are textual objects, related to different expression-substances, that our culture already presents as such, and are recognisable to us thanks to implicit shared criteria (communicativity, perceptible presence of limits, apparent functioning of an internal structure, etc.).

The euphoria that analysis of texts brings about represents a great step forward in semiotic research of the last decades of the 20th century, and it shows the fertility of a rigorous, but at the same time dynamic and adaptable, methodology. However, paradoxically, an important question is never mentioned: what are we talking about when we say 'text'? Are we sure we are all referring to the same theoretical construction? Or are we accepting different concepts of text coming alternatively from rhetoric, philology, linguistics, philosophy or literary criticism? The main model remains the linguistic one, especially in its literary manifestation, and that explains why the greatest quantity of textual examples come from literature (they are mainly poetic and narrative texts); the contribution of similar disciplines is also very important – film theory and communication research above all. Among the authors that, more than others, have tried to integrate practical analysis with a theoretical definition of textuality are Umberto Eco and Algirdas J. Greimas. I will now analyse their work in order to report their ideas but also to show which research pathways have originated from their studies.

6.1 Interpretative cooperation

In Eco's work, the textuality question is still intertwined with a more general reasoning about signs, though seen in the light of the principles of Charles S. Peirce's pragmatic philosophy (Eco 1979). According to Peirce, a sign is a dynamic reality, without any specific material or formal shape that triggers a virtually inexhaustible process of interpretation in which signifiers and signifieds recall one another. The signified of a sign is, for Peirce, another sign into which the first has to be translated, or at least approximately connected, in a virtually endless cross-reference game where any kind of expression-substance (gesture, verbal language, images, etc.) conducts interpretation (an idea that conciliates Derrida and Ricoeur). This goes on until some institutional stabilisation takes place, a kind of interpretive 'habitus' that crystallises the continuous escaping of interpretants. In a similar way, according to Eco (1979, 1990, 2002), the concept of interpretation as an inherent constitutive element of sign and text (the two of them sometimes differing just in terms of size) confirms the idea – so important for textual semiotics – that *the semantic plane prevails on the plane of expression*. The text, says Eco (1979), is an expanded lexeme, just like the lexeme is a concentrated text. The meaning expands and contracts in different forms of expression, from a single word to a whole work and, every time, it gives rise to more or less complex and fundamental interpretive activities. So, in Eco's view, to integrate the concept of text into semiotic analysis, though referring continuously to text linguistics, does not entail giving up the semantic analysis of terms. Terms, in fact, at their specific level, keep some literal meaning and, above all, presuppose discursive context and communicative situations within which they acquire their whole effective meaning that dictionaries record, institutionalising them.

A word, a phrase, a text or a set of texts triggers, through its internal semantic and syntactic structure, a fundamental pragmatic work of the addressee that is led to activate more or less large parts of his/her inner 'encyclopaedia' (that is to say his/her linguistic, lexical, textual, intertextual and generally cultural competences), in order to complete the meaning that the text leaves hanging over. In Eco's interpretive semiotics, the text looks like a kind of cultural configuration, a continuous dialogue between said and unsaid, between what it explicitly states and what it merely presupposes, promises, implicates and that the addressee must be able to grasp by means of his previous competences. The text is a *macchina pigra*, a lazy machine with many 'holes', interstices and empty spaces that the reader (since Eco takes into consideration written texts) has to fill with a meaning that is, at the same time, written and inferred, inside and outside, textual and cultural, objective and subjective (Eco 1979). On one hand the reader of the text, defined as the 'Model Reader', is part of the textual machine and takes on himself the task of repairing its innate laziness with a cognitive activity somewhat foreseen by the text; on the other hand, such textual pre-figuration of the reader acts in order to guide the empirical reader to understand not just what the text said but also, more importantly, what it presupposes – this, at least, is the utopian desire of the author, the expectation of a perfect realisation of the work.

From this derives a series of corollaries.

(i) First of all, there is the idea that not every text has the same degree of laziness and, consequently, not every reader is stimulated in the same way: for example, according to its communicative mandate, a pedagogical text explicates many more contents than an aesthetic one; a soap opera reveals more of its plot than a literary tale; a TV consumer programme gives a more detailed description of a product than an advertisement. It is important, though, not to mistake the degree of explicitness for the expressive size of a text and, consequently, for the quantity of information it contains. A soap opera discloses a great part of its plot, but this is very poor from the narrative point of view. An advertisement is very short but dense with information. The dialectic opposition between concentration and expansion is relevant once more and, since it depends on the interpretive activity of the addressee, it acquires a mainly pragmatic value.

(ii) Secondly, it appears clearly that the text Eco is referring to is a cerebral occurrence, so that the interpretive activity is an almost exclusively inferential, cognitive and quite logical work. The problem of the dialectic opposition between said and unsaid does not trigger, for example, passionate tensions – if they are not inserted in a more general will of understanding the meaning of the text – by anticipating narrative paths, making hypotheses on how the plot continues or guessing the end. The Model Reader's passions are mental emotions.

(iii) Thirdly, since the reader is inscribed into the text and their interpretive activity has been foreseen by the text itself, there is a precise difference between the actual *interpretation* of the text, activating in some way semantic paths already anticipated and written, and the *use* of the text, that is an idiosyncratic one and that derives from the text. looking for interpretive paths that are absent in the text as such. Here Eco is openly opposing deconstructionism, because it sees the text as a simple pre-text for any kind of speculation. *Foucault's Pendulum*, Eco's second novel, tells the fantastic (though plausible) story of the mess caused by unforeseen interpretations of certain texts and of the aberrant use of them leading to foolish, if not completely crazy, behaviours (see Marrone 2018).

(iv) Fourthly, it appears clearly that the dichotomy between incorrect uses and allowed interpretations of texts that could be read as prescriptive, takes place on the changing background of cultural contexts, of overall anthropological configurations, that do not establish any rule *a priori*, but present themselves as the theatre of continual *negotiations* between authors and readers, authors and authors, readers and readers, authoritative institutions and liberating trends – that is to say, basically, between texts and texts. The *encyclopaedia* that the Model Reader uses to make his/her interpretive hypothesis is actually the competence the text requires of its reader to be understood, and it also has a textual nature, or better, an intertextual, and therefore generally cultural, nature. Culture is a set of texts without any preset hierarchy, without any Porfirio's tree to determine absolute scales of logical or metaphysical values (Eco 2002). It is a dynamic place where continual negotiations take place between the actors of communication, in order to establish, every time in a different way, what is use and what is interpretation of a text, but more deeply – as we will see for Greimas and Lotman – what a text is and what it is not, in certain conditions and according to certain relevant elements. Hence what Eco himself calls *pansemioticism*, according to which the whole day-by-day life, not just culture, "appears as a textual net in which reasons and actions, expressions pronounced with communicative aims and the actions they trigger, become elements of a semiotic fabric in which each thing interprets another" (Eco 1979: 43).

6.2 Cultures and genres

A very important double radicalisation of interpretative semiotics is the theoretical project of François Rastier (2001), a Hjelmslev scholar and a student of Greimas's. He advocates a semantic of text that rigorously respects the limits of the semiotics of cultures. While, according to Eco, culture is an encyclopaedia or reservoir of memories that constitutes the background for the reconstruction of cognitive inferential processes, for Rastier it also includes social practices from which texts take not just their sense and value, but the reason for their existence. Against the cognitivist idea of a naturalisation of meaning (that Eco himself sustains, mainly in his late works), Rastier claims the opposite perspective, that is *a culturalisation of meaning*: a kind of general regulative hypothesis enclosing any description, analysis and theory of texts. According to Rastier, texts have three fundamental characteristics: (i) they are actively present in culture, documented, established, not artificially made up by the scholar as an *exemplum fictum* aimed at demonstrating some prior theory; (ii) they constitute themselves within a certain routine or social praxis; (iii) they are fixed to an empirically observable medium (such a medium could ideally be of any kind, but, again, it is thought mainly as a linguistic, verbal and, for the most part, written product). This leads to the radicalisation of the anthropological perspective and, at the same time, of the concept of interpretation. According to Rastier, interpretation is not a tool for assigning a meaning to a given expression anymore, but for recognising the expression itself. That is to say: the signifier supporting the signified. If we think of semiosis as the reciprocal presupposition of expression and contents (as Rastier thinks according to Hjelmslev, not to Peirce), interpretation is the variable item that acts as a base for the construction of that relationship of signification within a given culture. Consequently, the idea that the expression always comes before its contents is not true; often it happens the other way round: starting from some semantic expectations, some particular anticipations of meaning, a particular expression is identified as the base for a text, that is to say that the text is understood as a specific cultural entity. Rastier explains this concept by showing how a genre (as a kind of discourse or text) gives rise to expectations that gradually lead to the identification of what belongs to the order of text and what does not, of what the text's limits are and what its internal structure is. Genre predetermination – a prayer, a riddle, a tragedy, a novel, etc. – is a given semantic configuration that the text, through its expressive medium, specifies or, sometimes denies, failing the addressee's expectations. For example, in so-called neotelevision, a television flow divided into small parts (programmes) has no sense, because those parts, those programmes, are not perceived as actual texts (and therefore they are not) as they were in paleotelevision. Hence the disappearing or weakening of the initial theme, the creation of long programmes (including smaller ones) and of very small interruptions, etc. Within this discursive practice, then, is the whole programming that becomes the text. The boundaries between text and context are far from being given and unchangeable; in fact, genres and configurations modify them continually.

6.3 Generative path and projects of description

According to interpretative Semiotics, the text is a dynamic space where present rules and cultural encyclopaedia dialogue together in a hermeneutic dialectic relationship. On the other hand, Generative semiotics, with Hjelmslev's teachings in mind, conceives textuality as an empirical manifestation of an underlying discourse, as the natural unit of every single human and social signification, regardless of its expression-substance or its interpretation. In Algirdas J. Greimas's theory (see Greimas & Courtés 1979, 1986), texts (of any kind and dimension) are the object of knowledge of semiotics which has to explain their conditions of possibility, their requisite of existence, their transformation laws. The reason for this is quite evident: texts (not signs, codes, languages, etc.) are formal filters through which humankind – groups, societies, cultures, etc. – can get access to a prior sense that humankind seeks to understand in order to give a meaning to its own cultural, social and political existence, often changing it in the process. Semiotics has to re-build the textual grid, bringing it back to life if it had been hidden or removed, or unveiling it if it was implicit and unknown. Greimas takes, as a starting point, Hjelmslev's unfinished project of an analysis of the plane of contents to be performed with the same method adopted for the analysis of the plane of expression. As in linguistics, phonemes are basic units made up of a small number of traits; so meaning is, for generative semiotics, an entity that can be divided into minimum elements. There is necessarily a difference, though: on the plane of linguistic expression every language produces its set of phonemes, by combination and generation, from a relatively small stock of acoustic traits. On the plane of contents, by contrast, things have to work differently (more so if we extend it from verbal languages to any kind of signification system), since setting beforehand the fundamental categories of 'sayable' and 'significable' that are the logical atoms of any past, present and future discourse and of any form of social and human communication, would bring semiotics back to a prescriptive metaphysics. Therefore, according to Greimas, if a rigorous structural semantic is to be founded, the idea of a general semantic field where universal and necessary architectures of communication can be traced, must be abandoned. Attention must be directed, instead, to local domains of meanings, bounded to a particular point in time and space and determined in different ways by different cultures in different historical periods. However, such closed semantic micro-universes (such as a corpus of fables, a mythological set, a literary tale, a dictionary definition), do not exhaust in themselves their field of action because, having as a starting point explicit criteria of relevance that allow the comparison among them, they show, at their deepest level, quite similar formal traits.

Hence comes the central position of text. If, on the one hand, coherent classifications are to be built to account for genres and species of semantic micro-universes by creating a formal typology of social discourses, on the other hand, it is also necessary to investigate the structures linking together those discourses from the point of view of deep structures. This does not happen through the combination of minimum elements – as according to componential semantics, based on a theory of codes – but through overall articulations of configurations of meaning, structured along different levels of depth and complexity. For Greimas, then, it is possible to find, underneath specific discursive universes in the form of texts, a deeper level from which they are generated. This, however, does not bring him back to universalistic or metaphysic hypothesis.

When he moves from general theoretical problems to description methods of semantic universes, Greimas moves his focus from the combination of simple elements to the identification of complex routes through which signification displays itself. The semantic scholar eventually meets Propp's *Morphology of the Folktale* (1958), to which Levi-Strauss (1963) also had called scholars' attention. Greimas (1983a) worked on the typology of spheres of action and on the set of narrative functions the Russian folklorist had identified, and thus he conceived the hypothesis that possible semantic universals can be found in wider transphrastic structures, something immanent that, for this reason, is able to support the circulation of discourses and their expressive manifestation. "The generation of signification", Greimas (1987c) wrote, "does not pass through, first of all, the production of utterances and their combination into discourse; it is relayed, in its trajectory, by the narrative structures and it is these that produce meaningful discourse articulated by means of utterances". Moving beyond sentence boundaries does not lead, as in other linguistic theories, to a generic text imagined as a superior level where given utterances are linked, but to the construction of an actual *narrative grammar* that relies on some basic structures of signification (the semiotic square) and changes the initial semantic project to wider semiotic research. Within a theory of signification conceived in such a way, narrativity ceases to be the general model on which any narrative is based, as it was in narratology; it becomes, instead, an *interpretive hypothesis* that makes the scholar able to explain any semiotic phenomenon, that is, any cultural data, both narrative and non-narrative.

The text appears, therefore, as the tip of the iceberg of the *generative path of meaning*, the place where this trajectory acquires an expression-substance and thus takes on an empirical concreteness, becomes communicable, cognisable and reachable (Greimas 1983b; Greimas and Courtés 1979). The generative path of meaning, however, is, in turn, the simulation of the different levels of relevance in which meaning textualises itself, expresses itself through a specific concretion. That is to say, any human and social signification can be described by the semiotician – and, before that, understood by social subjects – at different levels of relevance, more or less abstract, more or less simple. Such a description can be performed at the level of elementary structures, where from a stock of relationships and transformation operations springs the first glimmer of meaning (the semiotic square); or it can be performed at the level of narrativity, where basic relationships assume anthropomorphic characteristics and give rise to tensions and clashes among the forces in play, while operations bring about identity changes. It can also be performed at the discursive level, where an enunciating subject uses the underlying structures by giving them specific actors, spaces and times and inserting them into specific themes and figurations. In this view, the text results from a different operation: textualisation. Textualisation works by stopping the generative path at some level and revealing it. It can be the square level, the discourse level or the narrative level, each with all its specifications.

Hence Greimas's well-known statement 'Outside the text, no salvation!'. The semiotician finds no salvation, because outside the text s/he would have to work on objects that are as entrenched in the linguistic and philosophical tradition in a manner embarrassing for a scientific theory of signification: sign, language, sentence, code, practices and so on. The lay person finds no salvation since s/he finds in texts – given or not, thus defined or not – the anchorage for fixing a meaning to their life, a tool to build and modify their subjectivity, a project to be part of a meaningful world that is, for this very reason, easier to live in. So it is not possible to go outside the texts, as Derrida said, and that is lucky, added Greimas, since if we go outside the text we would find no salvation. In order to enter more deeply into the question, let us look at the textual place where Greimas uttered this slogan. It was in 1983 at the end of a conference dedicated to him, and when he was publicly questioned, he affirmed:

> The first education I was given was as a philologist, and thanks to an excellent teacher I claim I have been educated as an excellent philologist: and that is something! I mean, here, that I respect the text, the reference, and the ideas of others. This influence is

equally important for what concerns textual practices. The preliminary operations to any semiotic analysis are found in philology, in the philological preparation of the text. This is an inescapable implication. Whether we are historians, linguists or logicians, we must know what a text is: the text is the starting point and the anchorage point of all our vociferations. We could say that it justified and found them. Later on, in the process of description, we walk far from the text, but it remains our only connection with the reality we live in, which is different from mathematical reality, from natural reality and so on[4]

(Greimas 1987b: 302, my trans.)

Later, on the same occasion, he answered a question about phenomenology and said:

Often, I have had occasion to talk about the importance of figurative models in the building of theories of language. Let us take into consideration the game of chess: it is an example all great thinkers have used: Husserl, Saussure, Wittgenstein. In their works, language is constantly compared to the game of chess. As for myself, the figurative model leading my way is found in Merleau-Ponty's work: it is the cube. What is the cube? It is – in a transposition towards the geometry of image – what wax was for Descartes. You can look at it from any angle: it always has a different face, but the actual cube remains unchanged forever. Here is a good definition of discourse as an autonomous object – "outside the text, no salvation!". It is a definition that allows us to talk about discourse without taking into account the variables of sender and addressee. The text is always there, as the cube is. There is the textual or narrative structure as an invariant we can use as a base for our analysis. Unlike any way it has been considered before, this invariant cannot be attributed either to the subject of the speech or to the addressee (as, for example, in Jauss's aesthetics); it is not possible to attribute everything to the producer or to the reader. Between the two there is the object. Its role could be confused, but the semiotic object's existence cannot be prevented. This is the starting point that forced me to accept the concept of semiotic existence, similar to the reality of mathematical objects. I think semiotics can imagine the existence of those simulacra, of those constructions, of objects that can be defined semiotically and whose kind of existence allows, in other words, the resolution of the problem of being, the ontological problem. A very important thing.[5]

(Greimas 1987b: 311, my trans.)

---

4 Original version: "La première formation que j'ai reçue, c'est la formation de philologue; et grâce à un maître remarquable, je prétends avoir été formé en bon philologue: c'est déjà quelque chose! C'est-à-dire que j'ai le respect du texte, le respect de la référence, de la pensée d'autrui. Cette influence est importante également en ce qui concerne les pratiques textuelles. Le préalable de toute analyse sémiotique est la philologie, la préparation philologique du texte. C'est un sous-entendu incontournable. Il faut savoir ce que c'est qu'un texte, qu'on soit historien, linguiste ou logicien: le texte est le point de départ et le point d'ancrage de nos vociférations, si l'on peut dire, il les justifie et les fonde. Ensuite, lors de la description, on s'éloigne évidemment du texte, mais c'est le seul rapport que nous ayons avec notre réel, différent du réel mathématique, du réel naturel etc."

5 Original version: "En ce qui me concerne, le modè (Greimas 1987b: 302) le figuratif qui m'a guidé, je l'ai trouvé dans le premier ouvrage de Merleau-Ponty: c'est le cube. Qu'est-ce que

Greimas's position appears quite clearly: the text is the 'referent' of the semiotician, the specific object of her/his investigations, of her/his 'vociferations'. That does not mean, however, that the text is a given object, a commonplace referent as in Morris. The text, in fact, like mathematical objects, has to be built gradually, 'prepared' in the same way as philologists do when they pass from their 'witnesses', variable in time and space, to the invariable textual object patiently retraced. Such a preparation presents itself as the deletion of any productive or receptive external fact, as a textual closure in the sense of the Formalists or the New Critics. Such closure, though, according to Hjelmslev's principle of immanence, is not the ontological assumption of a natural fact, but a precise methodological strategy that makes it possible to pass – as in the famous cube of Merleau-Ponty – from multiple perceptions of a single experience of a fact to the cognitive determination of a whole. The paradoxical cube of the *Phenomenology of Perception* is not found in the immediate apperception, but in the cognitive synthesis that can be executed *a posteriori*, in such a way that it gradually *becomes* an object of possible perception; in the same way, the text is, for the semiotician, a paradoxical object: it is created as a starting point and the reason why we find it at the end of the route is that we are already bound to it from the beginning. Thanks to multiple levels of relevance helping to bring it into focus (i.e. the levels of the generative path), the text confirms itself, in the end, as the object of analysis from the beginning. This is the only reason why the text is 'at the end' of the path: it is at the end of the path – that is a route to knowledge not to creation – that expression finds its way out as an autonomous plane having led the way throughout the route to the process of retracing the semantic articulations of the

---

c'est que le cube? C'est un peu, dans une transposition vers la géométrie de l'image, la cire chez Descartes, je crois. Vous pouvez regarder de tous les cotés, c'est chaque fois une apparence différente, mais le cube, en tant que tel, reste identique de toute éternité. Voilà une bonne définition du discours en tant qu'objet autonome – « hors du texte pas de salut! ». C'est une définition qui nous permet de parler du discours indépendamment des variables que constituent l'émetteur et le récepteur. Il y a toujours le texte, comme le cube: il y a la structure textuelle ou narrative, comme un invariant sur lequel peuvent se fonder nos analyses. Il ne s'agit pas de réduire cet invariant, comme on le fait trop souvent, soit au sujet de l'énonciation soit de l'énonciataire, comme dans l'esthétique de Jauss par exemple: tout ne se ramène pas au producteur ou au lecteur. Non, entre les deux, il y a l'objet. On peut voiler son rôle mais il n'empêche que les objets sémiotiques existent: tel est le point de départ qui m'a obligé à mettre en place le concept d'existence sémiotique, un peu comme il y a la réalité des objets mathématiques. Je pense que la sémiotique peut imaginer l'existence de ces simulacres, de ces constructions, des objets qui peuvent être définis sémiotiquement et dont le type d'existence permet, autrement dit, d'évacuer le problème de l'être, les problèmes ontologiques. »" (Greimas 1987b: 311)

text. Consequently, for Greimas, contents do not come *first* and expression *second*, but the two planes build one another as Hjelmslev thought. There is a difference, though, between the linguist and the semiotician: for the first, contents are a base supporting the manifestation of articulations of the expressive significant; for the second (aiming at solving the problems of the 'specificity' of every single language), it is expression that is the base supporting the articulations of the cultural contents that, by definition, transcend differences between expressive substances and are continually translated from a sign system to another. Even when, according to the principle of non-conformity of planes, it is necessary to study the contents as such, it always happens on the background of an expression that supports them.

Hence a number of theoretical consequences arise. First of all, the object that semiotics places at its 'empirical level' – semiosis – it is not a fact in itself, but it is constituted *as if it was* a natural fact. Natural here means obvious, routine (Greimas 1970; see Marrone 2006, 2011). It has, as well, the sense linguists give to this term when they speak of 'natural languages'. Therefore the semiotic perspective is based on a double constitutive action: the object of knowledge is something that is 'given' as a starting point of immanent descriptions, and 'created' at the same time – a creation that must be justified and motivated at the level of method, of theory and of epistemology. It is not strange, then, that in his *Semiotics and Language: an Analytical Dictionary*, Greimas states clearly that every object-semiotic, that is "any manifested entity under study (starting point for the analysis), exists only within the framework of a descriptive project and therefore presupposes a metasemiotics that, theoretically, encompasses it" (Greimas & Courtés 1979, entry 'Semiotics'). The empirical experience of semiosis cannot be defined as given or natural, unless it is in the sense of what we are used to (as Peirce's habitus is what stops the escape of interpretants, the cultural assumption of a sign fact). To the entry 'Text' in the same book, Greimas addresses the question once again, stating that the semiotic dimensions to analyse, the object-semiotics, are and remain texts that are built according to the level of relevance one decides to take into consideration. Hence the idea that a text is made up only of those semiotic elements fitting the theoretical goal of the description: this project must be made explicit and justified (and for that it is 'theoretical') and cannot be subject to the arbitrary subjectivity of the researcher (remember the previous recall to philology) (Greimas & Courtés 1979, entry 'Text').

Consequently, the task of the semiotician is divided into four different levels – the empirical, the methodological, the theoretical and the epistemological – interconnected to each other and thus justifying one another. It is as if different 'souls' of research (empirical, methodological, etc.) walked alongside each other without any of them prevailing on the others. Semiotics, therefore, is not the application of an empirical fact, of a theory and of a methodology philosophically controlled. If, as I said, *the empirical experience of semiosis is built gradually*, starting from a theoretically and philosophically controlled methodology, we have to place the fields of action of semiotics into a circularity, or in a net, in order for the epistemological and the empirical experiences to be already and always in a relationship with each other. The upstream philosophical choices about the identification of the object of analysis allow its reconstruction as a text to analyse. So, for example, the philosophical decision about the nature of passion (something that does not oppose reason and action) makes it possible to analyse its textual dimension (Greimas and Fontanille 1993). Or the assumption *a priori* of the existence of a continuity between perception and meaning gives rise to the textualisation of aesthetics (Greimas 1987a). The *empirical level is not the first level of semiotics but it is considered as such through a previous operation of construction of the data*, that sometimes is forgotten, hidden or not justified enough, but it is always constitutive of semiosis. On the other hand, *the object of knowledge of semiotics*, at least from Saussure onward, *is not a thing nor a concept, it is just a relationship between the two*. It is, therefore, necessary to suppose the existence of some kind of element – cultural, historical, scientific, social – that is a constructing subject, either individual or collective, taking charge of placing the relationship, of making it relevant and valid within the socio-cultural universe.

This is the conceptual framework within which Greimas' other important statement has to be placed. It is the statement introducing his famous and extended analysis of Maupassant's *Deux Amis*, according to which the text is the 'savage' for the semiotician (Greimas 1976): that is to say, the text is not an obscure and unintelligible fact which has to be faced one way or another, but the result of an ongoing *negotiation* between pre-set models and cultural resistances to analysis (see Fabbri 2007; Cavicchioli 2004). As the 'savage' is the condition and the incentive for the ethnologist to question his own conceptual and mental categories, so the text is the incentive for the semiotician to get a deeper knowledge of his own models of analysis. And as the 'savage' is not the 'good savage' – i.e. pure and uncontaminated humanity, the natural condition of human existence – but a subjectivity that has been produced and raised within specific, other cultural conditions, *the text*, as well, *is never innocent*, it is never a pure element but a hypothesis for the description of signification. The whole of Greimas's (1988) work about *Deux Amis* confirms this: it does not present a critical reading of the short story, nor an interpretation of an aesthetic evaluation; it is a completely different operation: the description of the internal structuring of the text's signification, the identification of its conditions of possibility as a semiotic 'dimension', a unit of meaning. At the end of the book, as Ricoeur

(1984–1988), says, emerges an original vision of the sense of the short story, as if it was some kind of hermeneutic reading of itself – but this is just a side effect of the analysis and not its main goal. Someone could certainly say that *Deux Amis* is a semiotic object that is, of course, a text, something that our culture defines without doubt as a literary work. But this also is a side-effect of the analysis. This appears clearly in many other analyses done by Greimas (and by his disciples afterwards) taking into consideration pragmatic situations, affective conditions intrinsic to the meaning of lexemes, whole mythological narratives and of course images, gestures, poetry and so on.

Thus, semiotic epistemology is for Greimas – as for hermeneutics and American textualism – basically constructivist, and it cannot be otherwise. This is one of the main differences between the science of signification and other human sciences: though the latter always discuss methodological problems, they never question the issue of data construction and they end up convincing themselves of the existence of an empirical world that is external and objective and that can be known regardless of the hypothesis of description, of the presupposed knowledge, of pre-set methods on which such a hypothesis is built. The question of method that semiotics shares with other human and social sciences, must not be mistaken for that of the construction of the object of knowledge, of the 'preparation of the text'. This question semiotics (corresponding with some currents of philosophy and sociology of science) deals with as a constitutive part of its research, offers it to other human and social sciences, but also to the so-called natural sciences, in order to achieve a greater epistemological awareness.

Once again, we need to go back to object-texts and model-texts. Basically, *all texts are constructed, but some of them, that forget or hide the work that has lead to their production, become objects, ontological facts*. Others keep the consciousness of their process of construction and can be used to create new textual realities. So there is no 'actual' text, on the one side, and other things that we 'metaphorically' thought of as texts, on the other. If we could, for a moment, leave behind our specific cultural condition, we could more easily admit that things are different: in general, *there are different textual formal grids for every culture* that allow the creation and management of human and social meanings. Every culture considers some of those as 'natural' texts and it applies to them specific 'recognition marks' that institutionalised their meaningful value. In our culture, for example, those recognition marks are book covers, title pages, frames, opening sequences, credits, theatre curtains and everything that marks discontinuity between inside and outside the text. What we are inclined to thinks of as 'actual' texts – being ourselves semiotic human beings in the first place – is what our culture wants and could build as such. An example is the construction of the book as a text-object that required many centuries of work (Genette 1997b) and that nowadays we often forget, as we think about the book as an obvious tool for the transmission of knowledge. Another good example is the picture as an artistic object enclosed in a frame (Stoikita 1999) that underwent the same oblivion and naturalisation as the image in a frame. The construction operations have been mainly forgotten – in the same way that truths in Nietzsche (1973) are metaphors that forgot the rhetorical artifice involved in their creation – so that we are inclined to consider as 'actual' texts those that have 'correct' signals pointing them out as texts. There would be, then, a previous *hidden textualisation* that produces a further *text neutralisation* giving rise to the differences between texts and non-texts: texts and experiences, texts and practices, texts and objects, scenes, situations, strategies and ways of life (see Fontanille 2008).

Therefore every object-text is actually a text built on previous cultural models (that is to say other texts). The only difference we can propose is the one between *silent cultural productions of texts* (given within every culture and based on implicit forms of giving and taking off of value to terms such as 'us' and 'them', to 'here' and 'elsewhere', to 'culture' and 'nature', etc.) and *explicit semiotic production of texts* (aiming at a metalinguistic knowledge, being themselves cultural operations bound to a specific goal of value). This is the reason why Greimas and Courtés (1982) make an important change: he substitutes the dichotomy 'natural *vs.* artificial' with the dichotomy 'scientific semiotic systems *vs.* non-scientific semiotic systems', where the first are "object-semiotics treated within the framework of a semiotic theory" and the others are "dimensions of the human and social world constructed implicitly" and so regarded as 'natural'.

7 Basics of sociosemiotics

It should be clearer now, or at least more explicit, the reason why, despite what many people think, *there is no opposition between sociosemiotics and textual semiotics for what concerns objects and methods*. On the contrary, these two perspectives integrate each other: while sociosemiotics deals with the social aspects of texts, textual semiotics studies the textual dimension of the social system. Vice versa, sociosemiotics analyses the conditions of possibility and exercise of the social system finding in them an intrinsic textual, narrative and discursive dimension, and textual semiotics retraces the conditions of possibility and exercise of texts, finding in them an intrinsic social, strategic and pragmatic, generally cultural dimension. Once again, *there is no opposition between the semiotic texts and the social contexts*: they have the same double nature and it is just the description project of the aware scholar that decides, each time, what is one and what is the other, what is relevant to the analysis (and is a text) and what is not (and is a context).

It is not unrelated that in the early 1970s the concept of text was seen as the antidote to sociological methodologies for the study of mass culture. In a fundamental proposal for the theoretical construction of sociosemiotics, Fabbri (2018) opposes to the so-called 'content analysis' that sociologists use in their description of mass media, a semiotic analysis of the texts circulating within the mass culture. While the first one "reveals a pre-Saussurean epistemology" (Fabbri 2018: 26) considering the message as a stock of lexical entities, of words bearing single meanings, the second one sees it as a configuration of meaning supported by deep semantic structures. The analysis of the semantic content conveyed by mass media is vital to explain and understand the culture they produce and they are produced by. Such semantic content, though, cannot be explained as if it was patent, simple and immediately understandable. The audience, in fact, can perceive it only partially depending on their particular cultural competences; it is delivered through different expression-substances; it takes a textual form that is more than the words, images or melodies it is made of. By using the theoretical results of linguistic and semiotic analysis in media analysis, the content analysis approach is radically revised, and thus a new image of the text emerges that is no more a 'black box' of which to observe inward and outward streams, but a formal crystal of which is possible to analyse the internal and external formal functioning.

In a similar way, the key figures of 1980s and 1990s sociosemiotics, mainly Jean-Marie Floch and Eric Landowski, have immediately made clear which link exists between textual and social studies. Floch (2000) emphasised the determining value of Greimas's slogan 'outside the text no salvation!' for the analysis of advertising campaigns and marketing mechanisms, design objects and fashion strategies, spatial organization and proxemic structures, comics and Russian icons, architecture projects and artistic images. Landowski (1989: 298, my trans.) highlights, instead, that "the reality sociosemiotics takes as an object of study, that can be identified as the socially constructed conditions of the signification ability of our discourses and our actions, is to sociosemiotics nothing but another form of textuality". According to sociosemiotics, then, the text is not an object but a model.

It is thus possible to overcome criticism by analysts of contemporary media culture (mainly in the field of sociology and cultural studies), according to which, since it deals with texts, semiotics cannot account for complex social practices beyond the boundaries of textuality (as it was understood traditionally), that are fundamental for the production of meaning. An example is the practices of media consumption that, for De Certeau (1984), are productive of meaning, but they do not depend on any code, nor are they included in any text. This kind of criticism is inconsistent, because it still sees the text as an object with an institutional closure, not as a model of semiotic research that the actors of communication negotiate on every occasion (as in a conversation), or that analysts establish according to the kind of investigation they have to perform.

Once the difference between text and context is neutralised, a key notion has to be taken into consideration: the concept of *discourse,* that comes from linguistic studies but finds correspondences also in cultural studies. According to the theory of information, the only function of a text was to convey a number of contents. From my perspective, by contrast, the text also presents an image of the communicative situation in which it is placed and of its addresser and addressee, thus dictating the practical rules for its fruition. In other words, *every text develops a discourse, it becomes part of a generic model that, at the same time, transcends the text and is created partly by the text itself.* A movie trailer or an advertisement, for example, does not convey just a more or less hidden persuasive content; they create an actual communicative scenario that becomes part of a specific kind of discourse: the promotional discourse following the release of a new movie or product. Without this communicative frame it would be impossible to deeply understand the meaning of the text. From this point of view, to go back to the concept of context serves no purpose for the study. The discourse is at the same time a social and textual reality, culturally defined and semiotically articulated. As such, it could be considered, alongside the text, as the preferential object of sociosemiotic investigation.

Sociosemiotics was born within cultural studies proposing a strong conceptual structure that was able to explain and understand a great quantity of social phenomena, from food to television, from publicity to the internet, from politics to fashion, from architecture to journalism and design, etc. Sociosemiotic research models cannot be used just in social sciences since it is not, as Barthes thought, only a methodology for human and social sciences. Sociosemiotics, in fact, deals with the mechanisms of production and articulation of meaning, thus placing itself at another epistemological level than human and social sciences: the level of the *critical examination* (as understood by Kant) of those sciences; that means looking for the formal conditions of possibility of sociality as such. In the semiotic perspective, as mentioned above, the social dimension is not an empirical fact, by reference to which it may be possible to unveil hidden laws; it is, instead, a constructed *effect of meaning* which is important to study the procedures that brought it into being. Landowski (1986: 207 my trans.) writes:

In its own way, general semiotics never actually stopped dealing with reality and, *a fortiori*, with sociality, both conceived as effects of meaning. In short and deliberately naive terms, the great issue the sociosemiotic scholar has to face is to account for 'what we do' in order for sociality to exist for us as such: how do we construct its objects and how do we play our part within them as talking and acting subjects. The empirical object of sociosemiotics can be defined, in this sense, as the set of discourses participating in the construction and/or in the transformation of the conditions of interactions between (individual and collective) subjects.[6]

Therefore, in a study about political discourse, Landowski (1989) describes the concept of Public Opinion as a character that is placed in different kinds of narratives, depending on the cases. From this point of view, sociosemiotic analysis does not focus on the origin of public opinion or on its effects on political life, but it seeks to retrace the formal system supporting the many 'stories' in which public opinion is placed, the fictitious world where the dynamics of political life develop. Although, for many centuries now, public opinion has been perceived as obvious, so much so that often people speak in its name, claiming its reasons, it is actually a semiotic construction as imaginary as effective, of which it is necessary to find the conditions of functioning before studying its ideological effects. According to Landowski, to question its possible manipulations or claim objective criteria for its analysis could deny to the scholar the possibility of understanding that its existence is due to a deeper manipulation, of which sociosemiotics is able to retrace the semantic, syntactic and pragmatic procedures. Similarly, instead of looking for the ways in which publicity persuades consumers to buy a product, Floch (2001) built a general model (the so-called axiology of consumption values) that was coherent and interdefined, thus allowing him to retrace the system of consumption choices that the advertising texts represent. The results of this analysis show that instead of choosing a given product for economic reasons, consumers attribute to it a set of values, thus projecting on their allegedly rational decisions, their own vision of the world, whose logic of functioning is one of Floch's objects of study. It could be a *practical* logic (when the object is advertised for its practical uses) or a *critical* logic (when economic ideas

---

6 Original version: "A sa façon, la sémiotique générale n'a cessé, dès le départ, de s'occuper du réel et, a fortiori, du social, conçus comme effets de sens. Formulée en termes succincts et volontairement naïfs, la grande question posée au sociosémioticien serait alors de rendre compte de 'ce que nous faisons' pour que le social (ou le politique, etc.) existe en tant que tel pour nous: comment nous en construisons les objets et comment nous nous y inscrivons en tant que sujets parlants et agissants. L'objet empirique de la socio-sémiotique se définit en ce cas comme l'ensemble des discours et des pratiques intervenant dans la constitution d'interactions entre sujets (individuels et collectifs)".

of cost effectiveness come into play), but it could also be a *utopian* logic(when the object becomes necessary for the realisation of the subject's identity) or *playful* logic (when the fact that the object is beauty or nice to play with overcomes the attention for its use). Therefore, sociosemiotic research looks further backward than sociology: while the latter pays attention to the empirical phenomena it sees in collective forms of life, the first one aims at retracing the procedures of meaning that bring about sociality, making of life a set of empirical institutional and collective phenomena. For semiotics, in sociality there is nothing patent or immediate but the very fact that sociality itself builds its own patency and immediateness, presenting as obvious and 'natural' what is actually the manifested result of immanent processes of signification.

8 Ethnology and semiotics of culture

We have mentioned above the very tight link existing between sociosemiotics and cultural studies; we also remarked on the importance of cultural assets in the constitution of reality and, on the other hand, the importance of textual mechanisms to the formation of different cultures. In order to examine this question in more depth, we have to look at the rich tradition of semiotic studies of culture, analysing the works of such authors as Claude Lévi-Strauss and Clifford J. Geertz and getting to the fundamental research of Jurij M. Lotman and his school. This trend, by combining sociosemiotics and anthropological semiotics, gives light to the possibility of studying cultural phenomena as texts. Actually, there are semiotics scholars that are used to define this hypothesis as 'ethnosemiotics' (Marsciani 2007).

In the anthropological field we come across the contrast between structural analysis and hermeneutic interpretation again. On the one side, we have Lévi-Strauss, promoting a formal and complete study of the internal functioning of cultural phenomena such as totemism, myth or kinship. On the other side, there is Geertz who criticises Lévi-Strauss's alleged formalism and proposes a more dynamic vision of cultural facts and of their analysis, in terms of a series of interpretive chains interrelating on the basis of apparently limited events such as the famous Balinese cockfight. If we look at the works of the two scholars in the light of the concept of text – that Geertz mentions explicitly, while Lévi-Strauss does not – we will see how the difference between them (and the trends they represent) is much smaller than it looks at first sight. The analysis of Lotman's works, in turn, will help us understand how the author, beyond the immediate evidence of his writing, is extremely helpful for a further analysis and re-structuring of many issues that have arisen till now.

8.1 Structural anthropology

The great and problematic figure of Claude Lévi-Strauss appears, at first sight, quite unrelated to sociosemiotic interests, particularly to textuality. The scholar's theoretical horizon is anthropology as the systemic analysis of cultural facts in general, a much wider perspective than the study of mass culture or of consumer society. Moreover, the large configurations Lévi-Strauss analyses – such as Amerindian mythologies, kinship systems, totemism – cannot be considered as texts, at least as what philology, linguistics, literature, hermeneutics and, often, semiotics regard as texts: a written linguistic product, mainly literary and thus highly idiosyncratic. On a closer look, though, things appear different. Thanks to the very wideness of the investigative horizon of anthropology, the concept of text can be free from its more traditional interpretations.

This is due to three reasons. (i) The first one is that both narrative and non-narrative texts studied by anthropologists are ethnic and folkloristic elements. This means that they do not have a recognisable author, they are mainly collective texts, orally transmitted so as to build a kind of *langue* that is social and abstract at the same time (see Jakobson and Bogatyrëv 1927). This is very interesting for the semiotician who is looking for the functioning parameters of signification that include, going beyond them, interpersonal linguistic communication or writing/reading processes. (ii) The second reason is that, in mythology, for example, myths are not actual textual units, but macrotexts or, even better, they are migration processes from one text to another, translation streams, intertextual chains with no beginning nor end that can be identified as a discursive genre, nor as one occurrence. Myth, Lévi-Strauss (1963: 129) says, is the result of its transformations, the combination of its translations: "the unity of the myth is never more than tendential and projective and cannot reflect a state or a particular moment of the myth" (a principle Ricoeur doesn't seem to understand). This, too, is a typical element of the mediatic universe, where the outline of every single textual event always appears on the horizon of a richer stream of textual events, continually building and re-building themselves within frameworks of genre that, in turn, are continually re-written. (iii) The third reason why structural anthropology can widen the concept of text and make an important contribution to sociosemiotic studies is that anthropology does not perceive the boundaries between a narrative text and an artefact, between the topology of a place or the organization of a ritual, as very well defined as it could be in some other disciplinary perspectives. A legend or a ceramic pot, a myth or a mask can equally be witnesses of some cultural configuration. From this point of view, the famous analysis Lévi-Strauss (1963) made of the Bororo village, whose structure mirrors the social hierarchy of the inhabitants, is emblematic of a sociosemiotics of spatiality and of city organisation; and the same can be said of the study about Amerindian masks, that was very important for Floch (1995) in his analysis of trademarks.

In other words, for Lévi-Strauss the text has to be understood not as a more or less elaborated linguistic artefact, but as any other object of anthropological knowledge, because it is mainly a bridge between expressive emergences and products of meaning, but also between series of complex manifestations of the material culture and large configurations that Lévi-Strauss would call, with many ambiguities, spiritual. The reasons for this theoretical position are quite clear. In the first place, cultural objects are texts because, as linguistic texts, they are not naive empirical elements, but constructed data whose construction system has to be explained; they are semiotic facts whose conditions of existence have to be understood. In the second place, this building procedure walks the path of a basic misunderstanding and tries to solve it: the misunderstanding between the point of view (the code system) of the so-called indigenous and the point of view (the culture) of the researcher. The text is the result of a translation process whose rules, conditions of exercise and necessary negotiations have to be made evident.

According to Lévi-Strauss, however, and actually anticipating what Geertz will say later on, the deepest essence of the text is for it to be the tool of necessary mediations between the world of sensible facts and the actual knowledge, a filter through which the sensible becomes a tool for intelligibility and vice versa, as in Saussure's view. "The operations of the senses", the anthropologist writes in the 'Finale' of his *Mythologiques*,

> have, from the start, an intellectual aspect, and the external data, belonging to the categories of geology, botany, zoology, etc., are never apprehended intuitively in themselves, but always in the form of a text, produced through the joint action of the sense organs and the understanding. (Lévi-Strauss 1981: 402)

Lévi-Strauss (1966) accepted the teaching of phenomenology of perception (*The Savage Mind* is dedicated to Merleau-Ponty) but moves beyond it by proposing a possible analysis of sensible experience from a textual perspective. According to Floch (2000), the famous *bricolage* constituting the *science of the concrete* of the so-called primitives (recycling waste materials to create original objects) is nothing but a discursive procedure aiming at creating actual texts. And, in fact, contemporary advertising communication uses it widely in order to create a visual and aesthetic identity of brands.

8.2 Interpretative anthropology

In his ethnological work, Clifford Geertz (1973, 1983, 1988) makes explicit reference to semiotics, since he defines the concept of culture as a semiotic concept. According to Geertz, and in line with Weber's ideas, culture can be identified as a net of meanings humans produce and get trapped in: a series of systems comprised of connected interpretable signs. On further scrutiny, we notice that those signs can be interpreted at least twice. The first time, they are interpreted by a determined social actor, i.e. the 'indigenous', who lives in a particular culture understood as a net of meanings, and gives a social and cultural sense to things and events of the world and to others' behaviour. The second time, they are interpreted by anthropologists that seek to retrace the series of interconnected systems and thus acquire the ability to interpret those signs, actually interpreting what is already an interpretation. The resulting concept of culture is neither substantialist nor cognitivist, but extremely dynamic; there is no ontological data in it, but meanings of things and events; no raw objects, but processes of signification; that is to say textual constructions and chains of them.

Every culture, according to Geertz, creates the tools to understand itself, to make sense of itself and to build an image of itself or a series of such images. Texts are those tools, both imaginary and real ones: Geertz refers to them as *inventions*, not because they are false, but because they have been produced intentionally. They can be discourses, rituals, behaviours, affections, objects, and, above all, twines of them all. Texts circulate around culture. As "acted documents" they are ready to be repeatedly subject to interpretation whose repercussion constitutes, hermeneutically, the (in process, ever-changing and unstable) identity of social subjects. Balinese cockfights, to mention a famous example, are a text for this very reason: because the individuals that, in different ways, are involved in it, watch the ritual and take part in it, feel and 'discharge' emotions based on cultural scales of values and implicit social hierarchies. The same thing happens to the audience of a Shakespeare play or to the readers of a Dickens novel who, by interpreting those texts, understand themselves better, and also understand the cultural environment they live in, "colouring experience with the light they cast it in" (Geertz 1973: 55). Social subjects live their ethnic experiences thanks to the texts of culture, where they, in a way, find a place for themselves. They also give to their ethnic experience an interpretation in all the senses of the term – that is to say cognitive, theatrical or translational – in order to understand the meaning, to give life to some social position, to translate them into other texts. It is easy to see here, beyond Ricoeur's and Peirce's influence.

Anthropologists are not detached observers that notice the semiotic mechanisms of a given culture through a precise descriptive external metalanguage.

And they could never be, since they are social subjects themselves, placed in their own history and culture, and since they are, at the same time, ethnologists. They are, actually, involved in the culture they mean to describe; they have to sink into it, aware that they will never be completely absorbed by it. "Anthropologists don't study villages (tribes, towns, neighborhoods . . .), they study *in* villages" (Geertz 1973: 451). An anthropologist's view is, therefore, an external and internal view at the same time. They interpret cultural signs not at the level of the 'indigenous', but at a secondary level that is their own: they let their culture interact with the one they are analysing. They interpret interpretations trying to build what Geertz, following Ryle, calls *thick descriptions*, that are not *thin*, since they have been constituted as stratified hierarchies of meaningful structures. That is, once more, textual construction. A thick description is the laborious result of a long meta-interpretive work that eventually constitutes itself as a text. A text that is, on the one hand, a cultural grid to read cultural facts with multiple levels of meaning arranged in specific orders; and, on the other hand, in the traditional sense of the term, a written linguistic product. If ethnography means to write reports (texts used by Geertz as examples for the theorisation about his work), ethnology means to create works (that, as literary works, Geertz analyses in order to retrace the sense of the anthropological work of many scholars).

8.3 Cultural models

Jurij M. Lotman's concept of text is fundamental and problematic at the same time. It is fundamental because more than once, though not continuously, he sees it as the centre and main object of semiotics understood as the study of any meaningful human phenomenon. It is problematic for many reasons: first of all, because Lotman used to work with other scholars and it is often impossible, though actually useless, to tell his contribution apart from the others; secondly, due to the language he used, Russian, and the many existing translations that do not always agree; lastly, because the term *text* is used with different meanings, often opposed to each other and always sources of misunderstandings. Two of the fluctuations of meaning of the term *text* create a problem we need to face here. On the one hand, we have the fluctuation between *text as literary work* and *text as cultural product* in a wider sense, including, for example, a ritual or an everyday behaviour, an artefact or a picture. On the other hand, we have the tension between what is perceived as *a text within each culture* (that will be defined as *emic* with anthropological and linguistic nuances) and *what the scholar takes as text independently and sometimes in contrast with the first one* (defined as *etic*). Clearly, those two concepts do not correspond with each other and intertwine. Because, on the one hand, every scholar, though he tries to be objective, belongs to some culture (as Geertz showed) and tends to impose his own idea of textuality onto his own theoretical and methodological models; and, on the other hand, because not every culture considers the text in the same way as it is understood by many scholars we mentioned above (as written, verbal, closed product, etc.). Then – against the many discussions about the 'worrying' metaphorical use of the concept of text to study something that in itself is not metaphorical at all (a concrete practice, an experience, a part of space, etc.) – Lotman shows us how, in some period not so far from us, some kind of everyday behaviours were regarded as texts because they were aestheticised, or in some other periods, even certain cities were considered as textual forms: they were written and read as miniated codes. In other words, not always in the *emic* versions, the text is the object-book, while in *etic* versions it is a model of analysis: in some cultures the *emic* is also used as a model, so much so that it is not the book that becomes a 'metaphor' of the city but quite the opposite, it is the city that guarantees the understanding of sacred or poetical texts.

Thus, it is possible to explain the misunderstanding, often seen as the limit of some textual or 'textualist' methodology, for which the semiotic idea of text as a formal model of analysis is just a reformulation of the symbolist ideal (known and accounted for by Lotman) of the Book as image of the world, and it has, as such, a precise aesthetic-metaphysical perspective. The claim that the artistic text as "artistic work is a finite model of the infinite world", can actually give rise to many misunderstandings, since this concept can be read both in the *emic* sense (e.g. about Mallarmé) and in the *etic* sense (e.g. regarding a work by Pushkin as a typical example of romantic Russian culture). Other statements by Lotman also give rise to enunciative ambiguities, even if his work as a whole does not leave doubts about the difference of level between the two positions, though confirming semiotic interest for both. As always happens, the question is how problem of the *readability of the world* is set, whether within or outside a culture. This reveals, on the one side, a trend that considers the world as a text, and philological analysis as the specific tool for understanding it (that, from time to time, comes back to life through history, from Confucius to Russian folklore, from Galileo to the symbolists); and it reasserts, on the other side, the fact that it is semiotically possible to understand as text "in the broadest sense, any communication recorded in a given sign system" such that "we can talk about [. . .] a ballet, a play, a military parade and all the other sign systems of behaviour as a word processor, to the extent that we apply this term to a written poem or a painting" (Lotman and Uspenskij 1984: 43).

The starting point for Lotman's (1977) thought about the text is to be found in the legacy of Russian Formalism, on the background of Soviet Marxism dominating the cultural environment the scholar has been working in during the major part of his career. Lotman (1977) identifies 'text' with the internal structure of literary works that guarantees their expressivity through some necessary features such as delimitation and hierarchy of internal organization levels, both syntagmatic and paradigmatic. The text, as according to Russian Formalism, is the work objectified or, more precisely, the *possibility of an operative description of the work itself* that, if it is not actually scientific, is still discussible and sharable. The objectification of the work through its textual description, however, does not look just to the form of the more or less 'alienating process', as formulated by Sklovskij, Tomasevskij, Tynjanov and colleagues, but it is also enriched by the plane of contents that is what actually makes a text, as formulated by structural linguistics. By taking into account the plane of expression and the plane of contents at the same time, the textual object becomes more dynamic, since it is now open both to the scholar's study techniques and to the culturally and historically specific context in which it was born. The relativity of the text thus arises, and its conformation (Lotman here agrees with Greimas) is established according to the scholar's project of description: the same material as, for example, Sterne's *Sentimental Journey*, acquires different values and meanings when considered as a simple textual fragment, as an unfinished aesthetic project mechanically deteriorated, or as the poetical realisation of the author and therefore as an actual text (see Lotman et al. 1977). Whence derives the importance acquired by the definition of the text, based on the context. It would be inappropriate to regard the latter as a deep economic structure, establishing in a deterministic way, as with vulgar Marxism, the function of the spiritual work; it acts, instead, as a semiotic link between the poetic project of the author, the literary tradition, the norms of the genre, the common language, cultural beliefs, and everyday life. Sklovskij's notion of artistic *estrangement* can be understood as a deviation from perceptive habits, but also as a distortion of literary canons or as a deviation from standard linguistic rules (see Erlich 1954). It changes its meaning in any case, because what really determines what is being 'alienated' and its meaning, is just descriptive relevance and context. The constitutive tension between internal textual structure and external cultural structure (that can be seen as the dynamic perceptive relationship between a figure and its background) determines the meaning of the work, but, above all, the possibility of its description and of objectifying its intentionality thus.

Instead of adjusting texts to contexts by seeking in the latter the explanation to the former, Lotman tries to retrace very wide anthropological models through highly detailed textual analyses realised in the literary field as well as in religious, historiographical, folkloristic, cultural texts and so on. According to Lotman (1977), poetry is the textual field where, more than in other fields, the work on language transcends the immediate artistic intentionality and creates new life styles that, at the same time, work as aesthetic creations and social organisations. In this perspective, Pushkin's work, one of Lotman's main objects of study (see Lotman 2009), is not just a corpus of poetry where the author practices his art in composing and versifying. It is, instead, a place where it is possible to understand completely the whole romantic way of feeling and so the whole late 19th century period. Therefore, to analyse the structure of an artistic text means to explain how it becomes "the bearer of a specific thought, of an idea" and, at the same time, "how the text structure is related to this idea" (Lotman 1977: 33). In other words, according to Lotman (1977), the thought (content) and the text (form) saying it cannot be told apart because what is there is just an *internal structural organisation of the text producing a particular thought*, and, later on, detaching from it and interacting with it. The relationship between text and context, though asymmetric and conflicting, becomes dialectic. On the one hand, the historical time is identified with the 'noise' that, according to Information Theory, communicative entropy produces; on the other hand, the aesthetic side of the text is an additional signification that, by guaranteeing the informative effectiveness of the text, constitutes the general configuration of the historical period within which the text itself is produced and spread. In other words, it is not possible to have texts and poetry without a socio-cultural horizon within which to think them and use them, as well as it is not possible for socio-cultural horizons to exist without texts. Therefore, large cultural models can be retraced only if starting from seemingly unimportant micro-structures of single texts in a mechanism that, for Lotman, is not mechanistic nor deterministic but dynamic, unpredictable and changeable instead.

From this starting point derives Lotman's reflection about text and context, one of the key elements of his entire work. We find, on one side, internal hierarchical structures of textual products; on another side, various cultural phenomena that link to textual products and have analogous semiotic structures; on yet another side, there are the principles of internal description, specific of every culture, that are the bases for metatexts to weigh the value of any text and any cultural phenomenon. All these elements fit together thanks to mechanisms or cultural models that Lotman uses to describe how, by opposing each other, cultures build their own specific identity. Among those elements there is no ontological difference *a priori*; there is a variable hierarchy of relations: *what is text can become context in a different cultural perspective* (e.g. a poetic work becoming a model for aesthetic production); what is behaviour can be described and valued by internal cultural metatexts (e.g. declarations of poetics, tables of genres, avant-garde manifestos, but also manuals of etiquette or catechisms); what belongs to a genre can, in time, be re-thought in another (e.g. a prayer or a scientific essay become literary works); and so on. *The text of culture* is both (i) the single text a culture produces and (ii) its whole textuality: the two things having the same form and being a model for one another.

It is therefore clear why texts appear as a condensed program of the whole culture (Lotman 1990, 2009). A culture is a set of similar but different languages, asymmetric and conflicting with each other but living together thanks to a wider cultural mechanism that is itself a language, or maybe a text. This language has a normative nature and dictates, for example, whether words or images are more important, what is the reciprocal role of oral and written speech, how much real life counts compared to the work of art and how much things are compared to signs, etc. *Culture is a set of texts where the referent of the first becomes the signified of the second* and so on forever, and where metatexts – more or less prescriptive poetics – play the same role of cultural texts. Here comes the famous concept of the *semiosphere* that describes culture by recourse to the image of a living being with innumerable organs and functions, facing all the problems a living being faces when trying to adapt to a given environment and the internal changes taking place in this process. Thus, the life of the text is compared to that of a living organism: it is possible only if the text relates with other organisms, in highly conflicting fields where the daily fight for survival allows the continuous transformation of languages and cultures. Human culture is no longer understood – like the early structuralism – as a static system originating from the primary modelling system of the world due to linguistic verbalisation; it is, instead, the result of a dynamic interlace between language and world, culture and extra-culture: it is the semiotic act that produces language and, as a repercussion, world and culture, and, as a presupposition, nature. Says Lotman (2009: 17): "the space, laying outside of language, enters the sphere of language and is transformed into 'content' only as a constituent element of the dichotomy content/expression. To speak of unexpressed content is non-sensical". According to Lotman, the external referent as an inactive and meaningless object, does not exist. This does not mean that (social and natural) reality does not exist, but that it does not exists in itself, outside the language,that acquires it as a specific content of its own, giving it different structures every time. Consequently, the problem to face is not how to link language to the world, but how to re-establish continuously the world as the object of language and the world as an extra-linguistic reality. What is extra-linguistic reality? Lotman's answer is crystal clear: it is the content of another linguistic reality. So the reason why there are at least *two levels of objectivity* (the one internal and the other external to a given language and culture) is that there are always at least two languages, two cultures. The reason why, in any immediate experience we have, there is an immediate relationship between the language and the world (or between subject and object), is that we hypothesise the existence of one language that is, instead, on closer examination, the result of many intertwining different languages. In turn, the reality laying outside of language is the result of contexts expressed in other languages.

The main idea concerning us here is that every culture establishes internally what is text and what is not, what semiotic configurations acquire the status of text – and thus obtain value and meaning to the culture itself (they have a 'global meaning' for the addresser and the addressee and a 'global function' for the scholar studying the culture) – and what does not acquire such status. Hence the axiological oppositions between oral and written speeches, religious and scientific texts, life and writings and so on. Scholars, on their part, by focusing on the textual nature of some configuration in a particular culture, will find at the same time the anti-textual nature of others, the culturalisation (therefore the textualisation) of what in that culture is not text. For example, when still lives spread and begin to be seen as actual texts, things that before where regarded as lifeless elements of the world, change their status and come to be regarded as signs: there are no more bottles, jars, plums or pheasants, but their signs. The cultural relevance of the pictorial genre causes a change in how people perceive reality.

This leads once more to a realisation that, on further scrutiny, it is not scholars who, when studying the cultural world, take as text a 'non-text' and analyse it as *if it was* a text in order to get some results; it is culture itself that models the world, generating a meta-mechanism that shows its emergences as texts and non-texts. Those emergences are actually modelled, too, since they can be anti-texts, opposing texts, or plain non-texts, pure 'things'. 'Things', in other words, are the anti-model of the textual model; they are negative models, but still models. Cities can be designed in a certain period as rational models of the world, as the realisation of the illuminist utopia; or they can appear in other cultures as the anti-model, the place of disorder, chaos, anti-reason and non-culture. Basically, says Lotman (2005), each text spreads its aura of context and if it is taken away from its original context, it creates a new one. For example, if a crucifix is taken out of the church where it has the value of sacred object, it creates a new place, the museum, where it has the value of a work of art. Each text creates what is outside it, in function of the global meaning it bears, and it is in dialogue with its outside, as a building, that, by its style, brings sense to the others along the same road, by means of analogy or contrast.

Therefore we are again in front of Derrida's idea according to which it is impossible – theoretically and practically – to have an outside-text, and the link inside/outside is essentially paradoxical. Any time this distinction is produced, whatever it is called (text/context; culture/nature; town/country; us/others; etc.), it is denied at a higher level, the level of cultural meta-mechanisms. The cultural meta-mechanism, however, is actually what makes that distinction operative by creating it, in a way. The context is a text, nature is cultural, country is in town, we are the others. As long as a distinction is produced, a paradoxical *homeomorphism* is immediately generated between the whole and its parts. That is the reason why Lotman, across his entire work, insists on the phenomenon of the *text within the text* that is not an unusual case of the *mise en abyme*, more or less relevant from an aesthetic point of view, but a kind of essential mechanism of the semiosphere, the same that causes the semiosphere itself to be a unit in which many semiospheres are generated *ad infinitum*. We are again behold the idea of relevance, so that – for any specific culture as well as for textual analysis – the same object, e.g. *The Belkin Tales* by Pushkin, can be regarded as one text, as a set of texts or as a part of a larger text (the 19th century Russian tale genre) according to cases (Lotman et al. 1977).

9 Farewell to representation

What has been said above sheds a brighter light on what should be the base for any sociosemiotic perspective: the fact that the concept of representation is irrelevant both from the methodological and from the theoretical point of view. If, within the text, expression and contents are in reciprocal presupposition, and if text and context define each other both in social culture and in textual/socio/semiotic analysis, the idea that texts are not actual realities but, at most, a 'representation' of reality (if not a fake one, then one directed to a particular goal that could be expressive, aesthetic, political, ideological and so on) fades away. It is the same idea that lies at the base of statements saying that a novel 'represents' a certain portion of the world, a movie 'represents' a certain social reality, an advertisement a certain life-style, but without being any of those things; and it is this idea that leads to thinking that it is more productive to study 'actual reality' than to focus on textual forms that are just a 'representation' of that reality. As I have been saying from the beginning of this work, however, this idea, the inheritance of a positivistic epistemological position of naive realism, has already been left behind by the majority of philosophical – mainly phenomenological and hermeneutic – studies but also by ethno-anthropological and social researches, by sociology and ethnography of science including, of course, semiotics. Yet, for many obvious reasons, it is still alive and well: with the support of common sense, it comes back as long as the defence against it is not vociferous; sometimes it even appears in semioticians' writings and debates.

Let us go back once again. First of all, the text is not a representation of the world because, simply enough, it includes the world itself within its boundaries as a content and, at the same time, it is part of the world and acts in it as a social force. To study the text – but more than that, to read it and enjoy it – does not mean to focus just on formal-expressive surface elements speaking of an outside reality; it means, instead, to understand how contents and expressions are created together and act in society. That is the reason why, when Greimas needs to study certain social and cultural phenomena, he looks into exemplar texts – explaining, of course, the reasons for his choice. His book about Maupassant (Greimas 1988), for example, is not, or *not only*, about *Deux Amis* as a tale, but it is about everything that is narrated in it: the Franco-Prussian war, everyday life in late 19th century Paris, the French feeling of peace, how French people never take themselves seriously as opposed to German people who have a rigid morale, and so on. In more detail, when Greimas analyses the dialogues between the two main characters, rich with clichés and commonplace opinions, he does not mean to study literary dialoguebut everyday conversation in general. Similarly, when he examines the useless attempts of the enemy general to get the password for entering the city, he means to study rhetorical and hermeneutical techniques of manipulation, persuasion, interpretation and contra-interpretation, etc. Moreover, by analysing how Maupassant (following the realist directions of the impersonality of the author and the refusal of any internal point of view of the characters) accounts for the cognitive dimensions of the characters, that is never explicit but always inferable from their behaviours, Greimas tries to retrace a global (or a real) cognitive space. In this way, a so-called fictional tale becomes a tool that gives access to the understanding and explanation of a set of facts and phenomena it presents as its content but that, at the same time, transcend it. Similarly, in *De l'imperfection*, where he means to study in a semiotic sense the sensorial and perceptive processes of the body, Greimas (1987a) does not describe those processes directly (as a psychologist would do) but studies how they are narrated in some literary texts (by Tournier, Calvino, Rilke, Tanizaki, etc.). From those texts, or even from fragments of them, Greimas extracts *discursive models* that in his opinion can be considered as examples and lead to a generalisation: this can be verified by applying them to other phenomena and other texts.

Considering the matter in more detail, we realise that this proceeding is not only an idea of Greimas, nor specific to semiotics. Lotman, as we said, also shows how some (literary and non-literary) texts can have an exemplary role in helping to understand some historical period, some cultural asset (Pushkin and Leopardi *are* Romanticism). Geertz (1973) does the same thing when he describes Balinese culture by examining a long passage from a novel by the Danish author Hans Jacob Helms. It is also what historians do when they read Proust's work to understand late 19th century French society. It is what consumes sociologists when they look in Zola's novels for the social value of the first shopping centres. It is what scholars of conversation do when they try to retrace the laws of presupposition through Hemingway's famous dialogues. It is what urban sociologists do when they read De Lillo's novels in order to understand the sense of the American metropolis. It is what Marx basically did, when he looked into Balzac's work for the explanation of the social psychology of capitalism. And, let alone literature, it is the same thing archaeologists do when they progressively retrace from small fragments of objects, buildings or bodies, a whole culture or historical period: they just interpret and analyse texts. It is the only thing they can do.

Texts, then, are also, and above all, documents, sometimes monuments, regardless of – or in spite of – the intentions of their authors who still have to be considered in sociosemiotic analysis. When we study a set of advertising campaigns in order to understand how everyday life-style has changed from the advent of the mobile phone (or of any other modern technology: from GPS to iPod, from notebook to the web), we have to take into account what kind of advertisement they have, what its strategic goals are, what is the intentionality of the text, as Ricoeur would call it, that is completely patent and explicit (see Marrone 2007). If we take a literary passage, we have to consider that it is an aesthetic text (see Marrone 2009). If we chose a furniture object – a sofa, a lamp – we have to bear in mind its functional origin. And so on.

Text-documents have, among others, a very important characteristic: they are *attested*, they exist and can be experienced concretely regardless of the needs of the analysis. They were not produced to be analysed but for other goals, with other intentions that – once identified – do not represent a problem for the analysis. On the contrary, they are a warranty of 'naturalness', 'naivety', 'authoriality'. Other verbal texts and experiential circumstances (interviews, tests, focus groups, laboratory experiments) produced *ad hoc* for the analysis, lack that characteristic, both because their circumstances of enunciation are affected by the observational and scientific intentionality of the researcher and because they circulate in culture just as documents for research and so they have no social value but to be witnesses of academic working. Consequently, if we examine an advertisement and a casual interview on the streets, the first one appears to be more spontaneous and natural than the second, while the second is more factitious than the first. In other words, although from a social and communicative point of view the advertisement is constructed while the interview is instinctive, from the sociosemiotic research point of view it is exactly the opposite. Once more it is a question of *relevance*. This is the reason why the distinction between 'field' work and 'desk' work is no longer useful and appears to be just a strong hypostatisation of epistemological thoughts as commonplace, non-existent, basically positivist. Any serious social research is made up by both field and desk work: every field observation is textual analysis and/or presupposes it; every textual analysis is field work and/or presupposes it (Marrone 2001).

So, sociosemiotics can work both on 'text' and 'non-texts', (i) because the latter are also texts when they have some meaning, (ii) because the former already have, as their content, the world researchers mean to describe, (iii) because the one and the other are social actors acting in the world, often along with other social actors with their own pragmatic and passionate competences, cognitive abilities and referential values. On the other hand, analysing so-called 'practices' as if they were not texts but some other fact that is thought to be 'pure' and immediate, means to fall back into forms of textuality without identifying them, and consequently without being able to control them – with deleterious results at the level of knowledge.

Recommended bibliography

On sociosemiotics and the semiotics of practices: Basso-Fossali (ed.) 2006 (on the relation between texts and practices); Albèrgamo (ed.) 2014; Barthes 1972 (on contemporary mythologies); De Certeau 1984 (on everyday life); Eco 1994a (on mass culture); Fontanille 2006b, 2008; Giannitrapani & Mangiapane (eds.) 2018; Greimas 1990 (on the relations between semiotics and social sciences); Halliday 1978; Halliday & Hasan 1985; Jeanneret 2008 (on everyday life); Landowski 1986, 1989, 1997, 2004, 2005 (fundamentals of sociosemiotics); Latour 2005; Marrone 2001; 2017; Marrone & Mangano (eds.) 2018 (on animals in today's cultures); Mangiapane & Migliore (eds.) 2021 (on Europe identities); Oliveira 1997 (on showcases); Oliveira (ed.) 2013 (on Landowski theory); Pozzato 2012; Thibault 1991.
On semiotics and rhetoric: Bertrand 1999; Lozano 2012; μ Group 1970; Mangiapane 2018.
On the dialogue between semiotics and the sciences: Biglari & Roelhens (eds.) 2018, 2019; Fontanille & Zinna (eds.) 2019.
On intersemiotical and intercultural translation: Dusi & Nergaard (eds.) 2000; Fabbri 2001, 2017; Fontanille, Sonzogni & Troqque (eds.) 2016; Jakobson 1966; Sedda 2012, 2019; Steiner 1975; Torop 1995.
On literary text, poetics and structural philology: Avalle 1973; Bertrand 1985; Bertrand 2000 (an introduction to literary semiotics); Bertrand, Provenzano, & Stiénon (eds.) 2014; Coquet 1972; Corti 1997; Culler 1975; Doležel 1998 (on possible worlds); Dorra 2013; Fontanille 1999; Genette 1983, 1997a; Geninasca 1997; Greimas 1988; Greimas (ed.) 1972; Lotman 1977; Ouellet 1992, 2000; Quéré 1992; Riffaterre 1971, 1979; Segre 1979; Todorov 1973, 1977; Uspenskij 1973; Zilberberg 1988.

On scientific texts: Bastide 1990; Bastide & Myers 1992; Chatenet & Mattozzi (eds.) 2013; Dondero & Fontanille 2012; Hamon 1980; Latour 1988, 1999; Latour & Bastide 1986; Latour & Fabbri 1977; Latour & Woolgar 1979.
On philosophical texts: Bordron 1987, Marrone (ed.) 1994.
On historical discourse: Hartog 2015; Lozano 1987; Koselleck 2004; Lozano & Salerno (2021).
On legal texts: Arnaud 1973; Bassano (ed.) 2021; Jackson 1985; Jacquement 1996; Landowski 1989.
On media, Communication, and new media: Bolter & Grusin 1999; Boutaud 1998, 2015; Cosenza (ed.) 2003; Eugeni 2010; Fabbri 2018; Mangiapane 2018; Manovich 2001, 2013; Ong 1982; Peverini & Spalletta 2009; Pezzini (ed.) 2002; Polidoro (ed.) 2018; Volli 1997.
On cinema, television and audiovisual texts: Bettetini 1984; Blanco 2003; Branigan 1984; Casetti 1999, 2002; Eco 2018; Jost 1987; Metz 1974, 2011, 2016.
On advertising and brand: Ceriani 2001; Floch 2000, 2001; Marrone 2007; Marrone & Mangano 2015; Traini 2006; Umiker-Sebeok (ed.) 1987.
On political discourse: Bertrand 2007; Greimas 2017; Lakoff 1990; Pezzini 2001; Schapiro (ed.) 1984.
On religious discourse and biblical text: Bertetti (ed.) 2021; Chabrol & Marin 1974; Delorme 1991, 2006; Entrevernes 1977; Leone 2004, 2010; Panier 1984, 1991.
On fashion: Barthes 1983; Floch 2000; Greimas 2000.
On design, objects and technologies: Baudrillard 1986; Baudrillard 1976; Beyart 2012; Beyart 2015; Deni 2002; Douglas & Isherwood 1979; Fontanille & Zinna (eds.) 2005; Landowski & Marrone (eds.) 2001; Mangano 2011; Ong 1982; Semprini 1995.
On music, sound, rhythm: Barthes 1977; Brandt & Carmo (eds.) 2015; Ceriani 2000; Estay Stange 2014; Jacoviello 2013; Spaziante 2007; Tarasti 1979, 1984.
On taste, food and cuisine: Boutaud 2005; Giannitrapani & Puca (eds.) 2020; Greimas 1989; Landowski (ed.) 1998; Landowski & Fiorin (eds.) 1997; Marin 1986; Marrone 2016; Marrone (ed.) 2020; Stano (ed.) 2015, 2016; Ventura Bordenca 2020.
On gesture, language of signs, and the body: Armstrong 1995; Badir & Parret (eds.) 2001; Dorra 2013; Fontanille 2004, 2011; Kendon 2004; Klima & Bellugi (eds.) 1979; Marrone 2001, 2009; Marrone (ed.) 2019; Pavis & Villeneuve (eds.) 1993; Rector (ed.) 2002; Richie 1975; Ventura Bordenca 2020.
On semiotics and psychoanalysis: Arrivé 1992; Darrault-Harris & Klein 1993; Kristeva 1980; Lacan 1966; Nathan 2001.
On comics and graphic novel: Baeten & Frey 2014; Floch 1997.

References

Ablai, Driss & Dominique Ducard (eds.). 2010. *Vocabulaire des études sémiotiques et sémiologiques*. Paris: Champion.
Adam, Jean-Michel. 1984. *Le récit*. Paris: Puf.
Adam, Jean-Michel. 1990. *Éléments de linguistique textuelle*. Bruxelles: Mardaga.
Adam, Jean-Michel. 1992. *Les textes: types et prototypes*. Paris: Nathan.
Adam, Jean-Michel. 2000. *Textes et genres de discours*. Paris: Nathan.
Albèrgamo, Maria (ed.). 2014. *La transparencia engaña*. Madrid: Biblioteca Nueva.
Akrich, Madeleine & Bruno Latour. 1992. A Summary of a Convenient Vocabulary for the Semiotics of Human and Non-human Assemblies. In Bijker, Wiebe & John Law (eds.). *Shaping Technology – Building Society: Studies in Sociotechnical Change*. 259–264. Cambridge (Mass.): MIT Press.
Alonso, Juan. 2005. *Le discours de l'ETA*. Limoges: Lambert-Lucas.
Armstrong, David F. 1995. *Gesture and the Nature of Language*. New York: Cambridge University Press.
Arnaud, André-Jean. 1973. *Essai d'analyse structurale du Code Civil français*. Paris: L.G.D.J.
Arrivé, Michel. 1992. *Linguistics and Psychoanalysis*. Amsterdam-Philadelphia: John Benjamins.
Arrivé, Michel & Jean-Claude Coquet (eds.). 1987. *Sémiotique en jeu. A partir de l'oeuvre d'A. J. Greimas*. Paris-Amsterdam-Philadelphia: Hadès-Benjamins.
Augé, Marc. 1995a [1991]. *Non-Places: Introduction to an Anthropology of Surmodernity*. London-New York: Verso.
Augé, Marc. 1995b. *Sense of the Other: the Timeliness and Relevance of Anthropology*. Stanford (Calif.): Stanford U. P.
Austin, John. 1962. *How to Do Things with Words*. Oxford-New York: Oxford University Press.
Avalle, D'Arco S. 1973. *Principi di critica testuale*. Padova: Antenore.
Bachelard, Gaston. 2014 [1958]. *The Poetics of Space*. New York: Penguin Books.
Badir, Semir. 2014. *Epistemologie sémiotique. La théorie du langage de Louis Hjelmslev*. Paris: Champion.
Badir, Semir & Herman Parret (eds.). 2001. *Puissances de la voix*. Limoges: Pulim.
Badir, Semir & Nathalie Roelens (eds.). 2007. Intermédialité visuelle. [Special Issue]. *Visible* 3.
Baetens, Jan & Hugo Frey. 2014. *The Graphic Novel: an Introduction*. New York: Cambridge University Press.
Barros, Diana Luz Pessoa de. 1988. *Teoria do discurso: fundamentos semióticos*. São Paulo: Atual.
Barros, Diana Luz Pessoa de. 1990. *Teoria Semiótica do Texto*. São Paulo: Ática.
Barros, Diana Luz Pessoa de. 2017. Les études de société selon la perspective de la sémiotique greimassienne. In Thomas Broden & Stéphanie Walsh-Matthews (eds.). 2017a. A. J. Greimas – Life and Semiotics / La vie et la sémiotique d'A. J. Greimas. [Special Issue]. *Semiotica* 214. 373–392.
Barros, Diana Luz Pessoa de. 2019. Réflexions sémiotique sur l'énonciation. In Veronica Estay Stange, Pauline Hachette & Raphaël Horrein (eds.). *Sens à l'horizon. Hommage à Denis Bertrand*. 273–284. Limoges: Lambert Lucas.
Barthes, Roland. 1967 [1964]. *Elements of Semiology*, London, Jonathan Cape.
Barthes, Roland. 1972 [1957]. *Mythologies*. London: Jonathan Cape.

Barthes, Roland. 1975a [1966]. An Introduction to the Structural Analysis of Narrative. *New Literary History*, 6, 2. 237–272.
Barthes, Roland. 1975b [1970]. *S/Z*, New York, Hill and Wang.
Barthes, Roland. 1977. *Image, Music, Text*, London, Fontana Press.
Barthes, Roland. 1982a [1970]. *Empire of Signs*: New York: Hill and Wang.
Barthes, Roland. 1982b. *A Barthes Reader*. New York: Hill and Wang.
Barthes, Roland. 1983 [1967]. *The Fashion System*: New York: Hill and Wang.
Barthes, Roland. 1985 [1982]. *The Responsibility of Forms*. New York: Hill and Wang.
Barthes, Roland. 1986 [1984]. *The Rustle of Language*. New York: Hill and Wang.
Barthes, Roland. 1988 [1985]. *The Semiotic Challenge*, New York: Hill and Wang.
Barthes, Roland (ed.). 1966, L'analyse structural du récit. [Special Issue]. *Communications* 8.
Bassano, Giuditta (ed.). 2021. *Semiotica del diritto*. Milan: Mimesis.
Bastide, Roger (ed.). 1962. *Sens et usages du terme structure*. The Hague: Mouton.
Basso-Fossali, Pierluigi. 2017. La passion et la figurativité. Les deux tentations greimassiennes face à la profondeur. In Thomas Broden & Stéphanie Walsh-Matthews (eds.). 2017b. La sémiotique post-greimassienne / Semiotics Post-Greimas. [Special Issue]. *Semiotica* 219. 219–237.
Basso-Fossali, Pierluigi. 2018. *Vers une sémiotique écologique de la culture*. Limoges: Lambert-Lucas.
Basso-Fossali, Pierluigi & Maria Giulia Dondero. 2011. *Sémiotique de la photographie*. Limoges: Pulim.
Basso-Fossali, Pierluigi (ed.). 2006. Testo – Pratiche – Immanenza. [Special Issue]. *Semiotiche* 4.
Basso-Fossali, Pierluigi, Denis Bertrand & Alessandro Zinna (eds.). 2018. *Utopies et Formes de vie. Mythes, valeurs et matières. Hommage à Paolo Fabbri*. Toulouse: Cam/S.
Bastide, Françoise. 1990. The Iconography of Scientific Texts: Principle of Analysis. In Lynch, Michael & Steve Woolgar (ed.). *Representation in Scientific Practice*. 187–230. Cambridge, MA: MIT Press.
Bastide, Françoise & Greg Myers. 1992. A Night with Saturn. *Science, Technology, & Human Values*, 17, 3. 259–281.
Baudrillard, Jean. 1968. *Système des objets*. Paris: Gallimard
Baudrillard, Jean. 1976. *L'échange symbolique et la mort*. Paris: Gallimard.
Baudrillard, Jean. 1988. *Selected writings*. Stanford (Calif.): Stanford University Press.
Baudrillard, Jean. 1990 [1983]. *Fatal strategies*. London: Pluto.
Beividas, Waldir. 2017. *La sémiologie de Saussure et la sémiotique de Greimas comme épistémologie discursive: une troisiéme voie pour la connaissance*. Limoges: Lambert-Lucas.
Benveniste, Émile. 1971 [1966]. *Problems in General Linguistics*. Miami: Miami University Press.
Benveniste, Émile. 2015. *Langues, Cultures, Religions*. Laplantine, Chloé & Georges-Jean Pinault (eds.). Limoges: Lambert-Lucas.
Bertetti, Paolo (ed.). 2021. *Semiotica del testo biblico*. Milan: Mimesis.
Berthelot-Guiet, Karine & Jean-Jacques Boutaud (eds.). 2014. *Sémiotique. Mode d'emploi*. Lormont: Le bord de l'eau.
Bertrand, Denis. 1985. *L'espace et le sens. Germinal d'Emile Zola*. Paris-Amsterdam: Hadès-Benjamins.
Bertrand, Denis. 1999. *Parler pour convaincre*. Paris: Gallimard.
Bertrand, Denis. 2000. *Précis de sémiotique littéraire*. Paris: Nathan.

Bertrand, Denis. 2007. *Parler pour gagner. Sémiotique des discours de la campagne présidentielle de 2007*. Paris: SciencePo University Press.
Bertrand, Denis & Veronica Estay Stange. 2017. Transversalité du sens et relations interartistiques. In Thomas Broden & Stéphanie Walsh-Matthews (eds.). 2017b. La sémiotique post-greimassienne / Semiotics Post-Greimas. [Special Issue]. *Semiotica* 219. 315–333.
Bertrand, Denis, Jean-François Bordron, Ivan Darrault, & Jacques Fontanille (eds.). 2019. *Greimas aujourd'hui: l'avenir de la structure*. Paris: AFS éditions.
Bertrand, Jean-Pierre, François Provenzano & Valérie Stiénon (eds.). 2014. Literature and Semiotics: History and Epistemology. [Special Issue] *Signata*. 5.
Bettetini, Gianfranco. 1984. *Tempo del senso*. Milan: Bompiani.
Beyart, Anne. 2012. *Sémiotique du design*. Paris: Puf.
Beyart, Anne. 2015. *Sémiotiques des objets*. Liège: Presses universitaires.
Beyart-Geslin, Anne & Nanta Novello Paglianti (eds.). 2005. La diversité sensible. [Special Issue]. *Visible* 1.
Blanco, Desiderio. 2003. *Semiotica del texto fílmico*. Lima: Lima U.P.
Biglari, Amir (ed.).2014. *Entretiens sémiotiques*. Limoges. Lambert-Lucas.
Biglari, Amir & Nathalie Roelhens (eds.). 2018. *La sémiotique en interface*. Paris: Kimè.
Biglari, Amir & Nathalie Roelhens (eds.). 2019. *La sémiotique et son autre*. Paris: Kimè.
Bolter, Jay David & Richard Grusin. 1999. *Remediation: Understanding New Media*. Cambridge (Mass.): The MIT Press.
Bondì, Antonino. 2019. Penser les intensités des signes. Le devenir des structures, entre philosophie et anthropologie sémiotique. In Denis Bertrand, Jean-François Bordron, Ivan Darrault & Jacques Fontanille (eds.). *Greimas aujourd'hui: l'avenir de la structure*. 302–313. Paris: AFS éditions.
Bordron, Jean-François, 1987. *Descartes. Recherches sur les contraintes sémiotiques de la pensée discursive*. Paris: Puf.
Bordron, Jean-François. 2011. *L'iconicité et ses images*. Paris: Puf.
Bordron, Jean-François. 2017. La nature de la signification: idéalité et plurivocité. In Broden & Welsh (eds.) 2017b. 13–31.
Boudon, Raymond (ed.). 1968. *A quoi sert la notion de "structure"?*. Paris: Gallimard.
Bourdieu, Pierre. 1991 [1982]. *Language and Symbolic Power*. Cambridge: Polity Press.
Boutaud, Jean-Jacques. 1998. *Sémiotique et communication: du signe au sens*. Paris: L'Harmattan.
Boutaud, Jean-Jacques. 2005. *Le sens gourmand*. Paris: Rocher.
Boutaud, Jean-Jacques. 2015. *Sensible et communication: du cognitive au symbolique*. London: ISTE.
Brandt, Per Aage. 1992. *La charpente modale du sens*. Aarhus: Aarhus U.P.
Brandt, Per Aage. 2004. *Space, Domains, and Meaning*. Bern: Peter Lang.
Brandt, Per Aage. 2017. D'où vient le sens? Remarques sur la sémio-phénoménologie de Greimas. In Broden & Welsh (eds.) 2017b. 75–91.
Brandt, Per Aage & José Roberto do Carmo (eds.). 2015. Sémiotique de la musique / Music and Meaning. [Special Issue]. *Signata* 6.
Branigan E., 1984, *Point of View in Cinema. A Theory of Narration and Subjectivity in Classical Film*, New York, Mouton.
Bremond, Claude. 1973. *Logique du récit*. Paris: Seuil.

Broden, Thomas. 2009. The Phenomenological Turn in Recent Paris Semiotics. In John Deely & Leonard Sbrocchi (eds.). *Semiotics 2008. Proceedings of the 33rd Annual Meeting of the Semiotic Society of America*. Houston(TX). 16–19 october 2008. Ottawa: Legas. 573–583.

Broden, Thomas & Stéphanie Walsh-Matthews (eds.). 2017a. A. J. Greimas – Life and Semiotics / La vie et la sémiotique d'A. J. Greimas. [Special Issue]. *Semiotica* 214.

Broden, Thomas & Stéphanie Walsh-Matthews (eds.). 2017b. La sémiotique post-greimassienne / Semiotics Post-Greimas. [Special Issue]. *Semiotica* 219.

Brown, George & George Yule. 1983. *Discourse Analysis*, Cambridge (Mass.): Cambridge University Press.

Buttitta, Antonino. 1996. *Dei segni e dei miti*. Palermo: Sellerio.

Calabrese, Omar. 1985. *La macchina della pittura*. Rome-Bari: Laterza.

Calabrese, Omar. 1999. *Lezioni di semisimbolico*. Siena: Protagon.

Calame, Claude. 2000. *Le récit en Grèce ancienne*. Paris: Belin.

Calame, Claude. 2019. De la narratologie structurale à la pragmatique énonciative: formes poétiques grecques entre récit mythique et action rituelle. In Denis Bertrand, Jean-François Bordron, Ivan Darrault & Jacques Fontanille (eds.). *Greimas aujourd'hui: l'avenir de la structure*. 165–181. Paris: AFS éditions.

Calloud, Jean. 1976. *Structural Analysis of Narrative*. Philadelphia: Fortress Press.

Calvino, Italo, 2002. *Mr. Palomar*. London: Vintage.

Casetti, Francesco. 1999 [1985]. *Inside the Gaze: The Fiction Film and Its Spectator*. Chicago (Ill.): Indiana U. P.

Casetti, Francesco. 2002. *Communicative Negotiation in Cinema and Television*. Milan: Vita e Pensiero.

Cassirer, Ernst. 1923. *Philosophie der symbolischen Formen. Vol. 1: Die Sprache*. Oxford: Bruno Cassirer.

Cassirer, Ernst. 1945. Structuralism in Modern Linguistics. *Word*, 1:2. 99–120.

Cavicchioli, Sandra. 2004. *I sensi, lo spazio, gli umori*. Milan: Bompiani.

Ceriani, Giulia. 2000. *Du dispositif rythmique. Arguments pour une sémio-physique*. Paris: L'Harmattan.

Ceriani, Giulia. 2001. *Marketing moving: l'approccio semiotico*. Milan: Angeli.

Chabrol, Claude & Louis Marin (eds.). 1974. *Le récit évangélique*. Aubier Montaigne: Cerf.

Charles, Michel. 1995. *Introduction à l'étude des textes*. Paris: Seuil.

Chatenet, Ludovic & Alvise Mattozzi (eds.). 2013. Images & dispositifs de visualization scientifiques. [Special Issue]. *Visible* 10.

Cobley, Paul. 2014. *Narrative*. London: Routledge.

Colas-Blaise, Marion, Laurent Perrin & Gian Maria Tore (eds.). 2016. *L'énonciation aujourd'hui*. Limoges: Lambert-Lucas.

Coquet, Jean-Claude. 1972. *Sémiotique littéraire*. Tours: Mame.

Coquet, Jean-Claude. 1985. *Le discours et son sujet*. Paris: Klincksieck.

Coquet, Jean-Claude. 1997, *La quête du sens. Le langage en question*. Paris: Puf.

Coquet, Jean-Claude. 2007. *Physis et logos. Une phénoménologie du langage*. Paris: Presses universitaires de Vincennes.

Coquet, Jean-Claude (ed.). 1982. *Sémiotique. L'Ecole de Paris*. Paris: Hachette.

Corrain, Lucia. 1996. *Semiotica dell'invisibile*. Bologne: Esculapio.

Corrain, Lucia. 2016. *Il velo dell'arte*. Firenze: La casa Usher.

Corti, Maria. 1997. *Per una enciclopedia della comunicazione letteraria*. Milan: Bompiani.

Cosenza, Giovanna (ed.). 2003. Semiotics of New Media. [Special Issue]. *Versus* 94–96.

Courtés, Joseph. 1976. *Introduction à la sémiotique narrative et discursive*. Paris: Hachette.
Courtés, Joseph. 1986. *Le conte populaire. Poétique et mythologie*. Paris: Puf.
Courtés, Joseph. 1995. *Du lisible au visible*. Bruxelles: De Boeck.
Culler, Jonathan. 1975. *Structuralist Poetics*. London: Routledge.
Culler, Jonathan. 1981. *The Pursuit of Signes*. London: Routledge.
Damisch, Hubert. 1972. *Théorie du nuage. Pour une histoire de la peinture*. Paris: Seuil.
Damisch, Hubert. 1984. *Fenêtre jaune cadmium, ou Les dessous de la peinture*. Paris: Seuil.
Damisch, Hubert. 2012. *L'origine de la perspective*. Paris: ChampArt.
Darrault-Harris, Ivan & Jean-Pierre Klein. 1993. *Pour une psychiatrie de l'ellipse. Les aventures du sujet en création*. Paris: Puf.
Darrault-Harris, Ivan (ed.). 2016. Journée d'hommage à la mémoire d'A.J. Greimas. [Special Issue]. *Actes sémiotiques* 116.
Darrault-Harris, Ivan & Jacques Fontanille (eds.). 2008. *Les ages de la vie. Sémiotique de la culture et du temps*, Paris: Puf.
De Beaugrande, Robert & Wolfgang U. Dressler. 1981. *Introduction to Text Linguistics*. Tubingen: Nyemeier.
De Certeau, Michel. 1984 [1980]. *The Practice of Everyday Life*. University of California Press.
Deleuze Gilles & Félix Guattari. 1994 [1991]. *What is Philosophy?* New York: Columbia University Press.
Delorme, Jean. 1991. *Au risque de la parole: lire les Evangiles*. Paris: Seuil.
Delorme, Jean. 2006. *Parole et récit évangéliques: études sur l'Évangile de Marc*. Paris: Cerf.
Deni, Michela. 2002. *Oggetti in azione*. Milan: Angeli.
Derrida, Jacques. 1988 [1967]. *On Grammatology*. Baltimore: Johns Hopkins University Press.
Descola, Philippe. 2013 [2005]. *Beyond Nature and Culture*. Chicago (Ill.): Chicago University Press.
Didi-Huberman, Georges. 2005 [1990]. *Confronting Images: Questioning the Ends of a Certain History of Art*. University Park: Pennsylvania State University Press.
Doležel, Lubomir. 1998. *Heterocosmica. Fiction an Possible Worlds*. Baltimore: John Hopkins University Press.
Dondero, Maria Giulia. 2020. *The Language of Images. The Forms and the Forces*. Berlin: Springer.
Dondero, Maria Giulia & Jacques Fontanille. 2012. *Des images à problèmes. Le sens du visuel à l'épreuve de l'image scientifique*. Limoges: Pulim.
Dondero, Maria Giulia, Anne Beyart-Geslin & Audrey Moutat (eds.). 2017. *Les plis du visuel*. Limoges: Lambert-Lucas.
Dondero, Maria Giulia & Jean-Marie Klinkenberg (eds.). 2018. Greimas et la sémiotique de l'image. [Special Issue]. *La part de l'oeil* 32.
Dondero, Maria Giulia & Nanta Novello Paglianti (eds.) 2006. Syncretismes. [Special Issue]. *Visible* 2.
Dorra, Raùl. 2013. *La maison et l'escargot. Pour une sémiotique du corps*. Paris: Hermann.
Dorra, Raùl, María Isabel Filinich, Luisa Ruiz Moreno, Blanca Alberta Rodríguez Vázquez & María Luisa Solís Zepeda. 2019. Réflexions sur le principe de narrativité. In Denis Bertrand, Jean-François Bordron, Ivan Darrault & Jacques Fontanille (eds.). *Greimas aujourd'hui: l'avenir de la structure*. 101–109. Paris: AFS éditions.
Douglas, Mary. 1996. *Thought Styles*. London: Sage.
Douglas, Mary & Baron Isherwood. 1979. *The World of Goods*. New York: Basic Books.

Ducard, Dominique. 2017. Language and the Game of Chess. Saussure, Hjelmslev, Wittgenstein and Greimas. In Thomas Broden & Stéphanie Walsh-Matthews (eds.). 2017a. A. J. Greimas – Life and Semiotics / La vie et la sémiotique d'A. J. Greimas. [Special Issue]. *Semiotica* 214. 199–217.

Ducrot, Owald, Tzvetan Todorov, Dan Sperber, Moustafa Safouan & François Wahl. 1968, *Qu'est-ce que le structuralisme?*. Paris: Seuil.

Ducrot, Oswald & Jean-Marie Schaeffer. 1995. *Nouveau dictionnaire encyclopedique des sciences du langage*. Paris: Seuil.

Duranti, Alessandro (ed.). 2001. *Key Terms in Language and Culture*. London. Blackwell.

Duranti, Alessandro & Charles Goodwin (eds.). 1992. *Rethinking context. Language as an interactive phenomenon*. Cambridge University Press.

Dusi, Nicola & Siri Nergaard (eds.). La traduzione intersemiotica. [Special Issue]. *Versus* 85–87.

Eco, Umberto. 1976. *A Theory of Semiotics*. Bloomington: Indiana University Press.

Eco, Umberto. 1979. *The Role of the Reader*. Bloomington: Indiana University Press.

Eco, Umberto. 1994a [1964]. *Apocalypse Postponed*. Bloomington: Indiana University Press.

Eco, Umberto. 1994b [1990]. *The Limits of Interpretation*. Bloomington: Indiana University Press.

Eco, Umberto. 1995. *Six Walks in the Fictional Woods*. Cambridge (Mass.): Harvard University Press.

Eco, Umberto. 2002 [1984]. *Semiotics and the Philosophy of Language*. Bloomington: Indiana University Press.

Eco, Umberto. 2018. *Sulla televisione. Scritti 1956–2015*. Marrone, Gianfranco (ed.). Milan: La Nave di Teseo.

Elkins, James. 1999. *What Painting Is. How to Think about Oil Painting, Using the Language of Alchemy*. New York – London: Routledge.

Entrevernes, Groupe d'. 1977. *Signes et paraboles. Sémiotique et texte évangélique*. Paris: Seuil.

Entrevernes, Groupe d'. 1979. *Analyse sémiotique des textes*. Lyon: Presses Universitaires.

Erlich, Victor. 1954. *Russian Formalism*. The Hague: Mouton.

Estay Stange, Veronica. 2014. *Sens et musicalité*. Paris: Garnier.

Estay Stange, Veronica, Pauline Hachette & Raphaël Horrein (eds.). 2019. *Sens à l'horizon. Hommage à Denis Bertrand*. Limoges: Lambert Lucas.

Eugeni, Ruggero. 2010. *Semiotica dei media*. Rome: Carocci.

Fabbri, Paolo. 2001. *Elogio di Babele*. Rome: Meltemi.

Fabbri, Paolo. 2007 [1998]. *Le tournant sémiotique*. Paris: Hermès-Lavoisier.

Fabbri, Paolo. 2017. *L'efficacia semiotica*. Marrone, Gianfranco (ed.). Milan: Mimesis.

Fabbri, Paolo. 2018 [1973]. *Le comunicazioni di massa in Italia: sguardo semiotico e malocchio della sociologia*. Marrone, Gianfranco (ed.). Rome: Sossella.

Fabbri, Paolo. 2019. *Vedere ad arte*. Marrone, Gianfranco (ed.). Milan: Mimesis.

Fabbri, Paolo & Isabella Pezzini (eds.). Affettività e sistemi semiotici. Le passioni nel discorso. [Special Issue]. *Versus* 47/48.

Fabbri, Paolo & Gianfranco Marrone (eds.). 2000. *Semiotica in nuce. Vol. 1*. Rome: Meltemi.

Fabbri, Paolo & Gianfranco Marrone (eds.). 2001. *Semiotica in nuce. Vol. 2*. Rome: Meltemi.

Fabbri, Paolo & Dario Mangano (eds.). 2012. *La competenza semiotica*. Rome: Carocci.

Fabbri, Paolo & Tiziana Migliore (eds.). 2014. *Saussure e i suoi segni*. Rome: Aracne.

Ferrara, Alessandro (ed.). 1980. Speech Acts Theory: Ten Years Later. [Special Issue]. *Versus* 26–27.
Floch, Jean-Marie. 1985. *Petites mythologies de l'oeil et de l'esprit*. Paris-Amsterdam: Hadès-Benjamins.
Floch, Jean-Marie. 1997. *Une lecture de Tintin au Tibet*. Paris: Puf.
Floch, Jean-Marie. 2000 [1995]. *Visual Identities*. New York: Continuum International Publishing Group.
Floch, Jean-Marie. 2001 [1990]. *Semiotics, Marketing and Communication*. New York: Palmgrave MacMillan.
Florenskij, Pavel. 2002. *Beyond Vision: Essays on the Perception of Art*. London: Reaktion.
Flores, Roberto. 2017. Narration and the Experience of History. In Thomas Broden & Stéphanie Walsh-Matthews (eds.). 2017b. La sémiotique post-greimassienne / Semiotics Post-Greimas. [Special Issue]. *Semiotica* 219. 511–528.
Fiorin, Losé Luiz. 1996. *As astúcias da enunciação*. São Paulo: Atica.
Fontanille, Jacques. 1987. *Le savoir partagé*. Paris-Amsterdam: Hadès-Beniamins.
Fontanille, Jacques. 1989. *Les espaces subjectifs. Introduction à la sémiotique de l'observateur*. Paris: Hachette.
Fontanille, Jacques. 1993. Le schéma des passions. *Protée* XXI, 1.
Fontanille, Jacques. 1995. *Sémiotique du visible. Des modes de lumière*. Paris: Puf.
Fontanille, Jacques. 1999. *Sémiotique et littérature. Essais de méthode*. Paris: Puf
Fontanille, Jacques. 2004. *Soma & Sema. Figures du corps*. Paris: Maisonneuve et Larose.
Fontanille, Jacques. 2006a [1998]. *The Semiotic of Discourse*. New York: Peter Lang.
Fontanille, Jacques. 2006b. Pratiques sémiotiques: immanence et pertinence, efficience et optimisation. *Nouveaux actes sémiotiques*, 104–106.
Fontanille, Jacques. 2008. *Pratiques semiotiques*. Paris: Puf.
Fontanille, Jacques. 2011. *Corps et sens*. Paris: Puf.
Fontanille, Jacques. 2015. *Formes de vie*. Liège: Presses universitaires de Liège.
Fontanille, Jacques. 2017. La sémiotique de Greimas: un projet scientifique de long terme. In Thomas Broden & Stéphanie Walsh-Matthews (eds.). 2017a. A. J. Greimas – Life and Semiotics / La vie et la sémiotique d'A. J. Greimas. [Special Issue]. *Semiotica* 214. 91–110.
Fontanille, Jacques & Nicolas Couégnas. 2018. *Terres de sens*. Limoges: Pulim.
Fontanille, Jacques & Claude Zilberberg. 1998. *Tension et signification*. Liège: Mardaga.
Fontanille, Jacques, Marco Sonzogni, & Rovena Troque (eds.). 2016. Translating: Signs, Texts, Pratictices. [Special Issue]. *Signata* 7.
Fontanille, Jacques & Alessandro Zinna (eds.). 2005. *Les objets au quotidien*. Limoges: Pulim.
Fontanille, Jacques & Alessandro Zinna (eds.). 2019. Dialogue entre la sémiotique structurale et les sciences. Hommage à A. J. Greimas. [Special Issue]. *Langages* 2013.
Foucault, Michel. 1970 [1966]. *The Order of Things. An Archaeology of the Human Sciences*. New York: Pantheon books.
Foucault, Michel. 1972 [1969]. *The Archeology of Knowledge*. New York: Pantheon books.
Freedberg, David. 1989. *The Power of Images. Studies in the History and Theory of Response*. Chicago: Chicago University Press.
Frölicher, Peter, Georges Günther & Félix Thürlemann (eds.). 1990. *Espaces du texte. Spazi testuali. Texträume*. Neuchatel: à la Baconnière.
Geertz, Clifford. 1973. *The Interpretation of Cultures*. New York: Basic Book.
Geertz, Clifford. 1983. *Local Knowledge*, New York: Basic Book.
Geertz, Clifford. 1988. *Works and Lives*. Stantford University Press.

Gell, Alfred. 1998, *Art and Agency: an Anthropological Theory*. Oxford: Clarendon.
Genette, Gérard. 1983 [1972]. *Narrative Discourse: An Essay in Method*. Ithaca (N.Y.): Cornell University Press.
Genette, Gérard. 1997a [1982]. *Palimpsests: Literature in the Second Degree*. Lincoln: University of Nebraska Press.
Genette, Gérard. 1997b [1987]. *Paratexts: Thresholds of Interpretation*. New York: Cambridge University Press.
Geninasca Jacques. 1997. *La parole littéraire*. Paris: Puf.
Giannitrapani, Alice. 2013. *Introduzione alla semiotica dello spazio*. Rome: Carocci.
Giannitrapani, Alice & Davide Puca (eds.). 2020. *Forme della cucina siciliana. Esercizi di semiotica del gusto*. Milan: Meltemi.
Giannitrapani, Alice & Francesco Mangiapane (eds.). 2018. Animals in Law. [Special Issue]. *International Journal of the Semotics of Law*, 31, 3 https://link.springer.com/journal/11196/volumes-and-issues/31-3 (accessed on 10 may 2021).
Gimate-Welsh, Adrian (ed.). 2000. *Ensayos semióticos*. Puebla: Benemerita Universidad Autonoma de Puebla.
Goffman, Erving. 1967. *Interaction Ritual*. Garden City: Doubleday.
Goffman, Erving. 1969. *Strategic Interaction*. Philadelphia: Philadelphia University Press.
Goffman, Erving. 1971. *Relations in Public*. New York: Basic Books.
Goffman, Erving. 1981. *Forms of Talk*. Oxford: Blackwell.
Gombrich, Ernst. 1951. *Meditations on a Hobby Horse*. Phaidon: London.
Gombrich, Ernst. 1960. *Art and Illusion*. Princeton: Princeton University Press.
Gombrich, E.H. 1999. *The Uses of Images. Studies in the Social Function of Art and Visual Communication*. London: Phaidon.
Goodman, Nelson. 1968. *Languages of Art*. Indianapolis: Bobbs-Merrill.
Goodman, Nelson. 1978. *Ways of Worldmaking*. Indianapolis: Hackett.
Goody, Jack. 1977. *The Domestication of the Savage Mind*. Cambridge: Cambridge University Press.
Goody, Jack. 1987. *The Interface between the Written and the Oral*. Cambridge: Cambridge University Press.
Greimas Algirdas Julien. 1956, L'actualité du saussurisme (à l'occasion du 40e anniversaire de la pubblication du *Cours de Linguistique générale*). *Le français moderne* 3.
Greimas Algirdas Julien. 1983a [1966]. *Structural Semantics. An Attempt at a Method*. Lincoln: Nebraska University Press.
Greimas Algirdas Julien. 1983b. *Du sens II*. Paris: Seuil.
Greimas Algirdas Julien. 1984. Sémiotique plastique et sémiotique figurative. *Actes sémiotiques – Documents* VI, 60.
Greimas, Algirdas Julien. 1985. *Des dieux et des hommes*. Paris: Puf.
Greimas Algirdas Julien. 1987a. *De l'imperfection*. Périgueux: Pierre Fanlac.
Greimas Algirdas Julien. 1987b. Greimas mis à la question. In Michel Arrivé & Jean-Claude Coquet (eds) *Sémiotique en jeu*, Paris-Amsterdam-Philadelphia: Hadès-Benjamins.
Greimas Algirdas Julien. 1987c [1970]. *On Meaning*, Minneapolis: University of Minnesota Press.
Greimas Algirdas Julien. 1988 [1976]. *Maupassant: The Semiotics of Text*. Amsterdam and Philadelphia: John Benjamins.

Greimas, Algirdas Julien. 1989. Basil soup or the construction of an object of value. In *Paris School Semiotics, Vol. II: Practice*, Paul Perron & Frank Collins (eds.). Amsterdam: John Benjamins. 1–12.
Greimas, Algirdas Julien. 1990 [1976]. *The Social Sciences. A Semiotic View*. Minneapolis: University of Minnesota Press.
Greimas, Algirdas Julien. 2000. *La mode en 1830*. Thomas F. Broden & Françoise Ravaux-Kirkpatrick (eds.). Paris: Puf.
Greimas, Algirdas Julien. 2017. *Du sens en exil. Chroniques lithuaniennes*. Saulius Zukas & Kestutis Nastopka (eds.). Limoges: Lambert-Lucas.
Greimas, Algirdas Julien. & Jacques Fontanille. 1993a [1991]. *The Semiotics of Passions. From States of Affairs to States of Feelings*. Minneapolis: University of Minnesota Press.
Greimas, Algirdas J. & Jacques Fontanille. 1993b. Le beau geste. *RS/SI* 13.
Greimas, Algirdas Julien & François Rastier. 1968. The Interaction of Semiotic Constraints. *Yale French Studies* 41.
Greimas, Algirdas Julien (ed.). 1972. *Essais de sémiotique poétique*. Paris: Larousse.
Greimas, Algirdas Julien & Jacques Courtés (eds.). 1982 [1979]. *Semiotics and language: an analytical dictionary*. Bloomington: Indiana University Press.
Greimas, Algirdas Julien & Jacques Courtés (eds.). 1986. *Sémiotique. Dictionnaire raisonné de la théorie du langage. Tome II*. Paris: Hachette.
Greimas, Algirdas Julien & Eric Landowski (eds.). 1979. *Introduction à l'analyse du discourse en sciences sociales*, Paris: Hachette.
Gumperz, John J. 1982. *Discourse strategies*. Cambridge University Press.
Hall, Edward T. 1966. *The Hidden Dimension*. New York: Doubleday.
Halliday, Michael A. K. 1978. *Language a Social Semiotic. The Social Interpretation of Language and Meaning*. Baltimore: University Park Press.
Halliday, Michael A. K. & Ruqaya Hasan. 1985. *Language, Context, and Text: Aspects of Language in a Social-Semiotic Perspective*. Geelong: Deakin University.
Hammad, Manar. 2002. *The Privatisation of Space*. Lund: Lund University.
Hammad, Manar. 2006. *Lire l'espace, comprendre l'architecture*. Paris: Geuthner.
Hammad, Manar. 2015. *Sémiotiser l'espace, décrypter architecture et archéologie*. Paris: Geuthner.
Hammad, Manar. 2017. *Sémantique des institutions arabes (du Croire, du Pouvoir)*. Paris: Geuthner.
Hammad, Manar. 2018. *L'instauration de la monnaie épigraphique par les omeyyades*. Paris: Geuthner.
Hamon, Philippe. 1980. *Introduction à l'analyse du descriptif*. Paris: Hachette.
Hartog, François. 2015 [2002]. *Regimes of Historicity: Presentism and Experiences of Time*. New York: Columbia University Press.
Hébert, Louis. 2018. *Dictionnaire encyclopédique de sémiotique*. http://www.signosemio.com/documents/dictionnaire-semiotique-generale.pdf.
Hébert, Louis. 2020. *Cours de sémiotique Pour une sémiotique applicable*. Paris: Garnier.
Hebert, Louis & Lucie Guillemette (eds.). 2007. *Intertextualité, Interdiscursivité et Intermedialité*. Laval: Laval University Press.
Hénault, Anne. 1979. *Les enjeux de la sémiotique*. Paris: Puf.
Hénault, Anne. 1992. *Histoire de la sémiotique*. Paris: Puf.
Hénault, Anne (ed.). 2002. *Questions de sémiotique*. Paris: Puf.
Hénault, Anne (ed.). 2020. *Le sens, le sensible, le réel*. Paris: Sorbonne Université Presses.

Hjelmslev, Louis. 1959. *Essais linguistiques*. Copenhagen: Travaux de Cercle Linguistique de Copenhagen.
Hjelmslev, Louis. 1961. *Prolegomena to a Theory of Language*. Wisconsin: The University of Wisconsin Press.
Hjelmslev, Louis. 1975. *Résumé of a Theory of Language*. Copenhagen: Nordisk sprog-og Kulturfolag.
Hymes, Dell. 1981. *"In vain I tried to tell you"*. *Essays in Native American Ethnopoetics*. Philadelphia: University of Pennsylvania Press.
Ingold, Tim. 2013. *Making: Anthropology, Archaeology, Art and Architecture*. London: Routledge.
Jackson, Bernard S. 1985. *Semiotics and Legal Theory*. London: Routledge.
Jacoviello, Stefano. 2013. *La rivincita di Orfeo. Esperienza estetica e semiosi del discorso musicale*. Milan: Mimesis.
Jakobson, Roman. 1963. *Essais de linguistique générale*. Paris, Minuit.
Jakobson, Roman. 1976. *Six leçons sur le son et le sens*. Paris, Minuit.
Jakobson, Roman. 1979. Coup d'oeil sur le développement de la sémiotique. In Seymour Chatman, Umberto Eco & Jean-Marie Klinkenberg (eds.). *A Semiotic Landscape / Panorama sémiotique*. The Hague: Mouton. 3–18.
Jakobson, Roman & Bogatyrëv Petr. 1929. Die Folklore als eine besondere Form des Schaffens. *Donum Natalicium Schrijne*. Nijmegen-Utrecht: Dekker e van de Vegt.
Jakobson, Roman & Morris Halle. 1956. *Fundamentals of Language*. The Hague: Mouton.
Jacquemet, Marco. 1996. *Credibility in Court. Communicative Practices in the Camorra Trials*. Cambridge University Press.
Jawoski, Adam & Crispin Thurlow (eds.). 2010. *Semiotic Landcapes. Language, Image, Space*. London-New York: Continuum.
Jeanneret, Yves. 2008. *Penser la trivialité*. Paris: Hermès-Lavoisier.
Jost, François. 1987. *L'œil camera*. Lyon: Presses Universitaires.
Jullien François. 2000 [1995]. *Detour and Access: Strategies of Meaning in China and Greece*. New York: Zone Books.
Jullien François. 2004 [1986]. *A Treatise on Efficacy: between Western and Chinese Thinking*. Honolulu: University of Hawai'i Press
Kendon, Adam. 2004. *Gesture. Visible Action as Utterance*. Cambridge University Press.
Kersyté, Nijolé (ed.). 2009. En quête de Greimas. [Special Issue]. *Actes sémiotiques* 112.
Klima, Edward S. & Ursula Bellugi (eds.). 1979. *The Signs of Language*. Harvard University Press.
Klinkenberg, Jean-Marie. 1996. *Précis de sémiotique générale*. Bruxelles: De Boeck e Larcier.
Klinkenberg, Jean-Marie. 2019. Greimas et la sémiotique du monde naturel. In Denis Bertrand, Jean-François Bordron, Ivan Darrault & Jacques Fontanille (eds.). *Greimas aujourd'hui: l'avenir de la structure*. 34–45. Paris: AFS éditions.
Koselleck, Reinhart. 2004 [1979]. *Futures Past: On the Semantics of Historical Time*. New York: Columbia U.P.
Kristeva, Julia. 1980 [1969]. *Desire in Language: A Semiotic Approach to Literature and Art*. Oxford: Blackwell.
Lacan, Jacques. 1966. *Ecrits*. Paris: Seuil.
Lakoff, Robin. 1990. *Talking Power. The Politics of Language in Our Lives*. New York: Basic Books.

Landowski, Eric. 1986. Sociosémiotique. In *Sémiotique. Dictionnaire raisonné de la théorie du langage. Tome II*. Greimas, Algirdas Julien & Courtés, Jacques (eds.). Paris: Hachette.
Landowski, Eric. 1989. *La société réfléchie*. Paris: Seuil.
Landowski, Eric. 1997. *Presences de l'autre*. Paris: Puf.
Landowski, Eric. 2004. *Passions sans nom*. Paris: Puf.
Landowski, Eric. 2005. *Les interactions risquées*. Limoges: Pulim.
Landowski, Eric (ed.). 1993. Hommages à A.J. Greimas. [Special Issue]. *Nouveaux actes sémiotiques* 25.
Landowski, Eric (ed.). 1997. *Lire Greimas*. Limoges: Pulim.
Landowski Eric (ed.). 1998. Sémiotique gourmande. [Special Issue]. *Nouveaux actes sémiotiques* 55–56.
Landowski, Eric (ed.). 2017. A.J. Greimas. Sept lectures pour un centenaire. [Special Issue]. *Actes sémiotiques* 120.
Landowski, Eric, Raul Dorra & Ana Claudia Oliveira de (eds.). 1999. *Semiótica, estésis, estética*. São Paulo-Puebla: Educ-Uap.
Landowski, Eric & José Fiorin (eds.).1997. *O gosto da gente, o gosto das coisas. Abordagem semiótica*. São Paulo: Educ.
Landowski, Eric & Gianfranco Marrone (eds.). La société des objets. [Special Issue]. *Protée* 29, 1.
Landowski, Eric & Ana Claudia Oliveira de (eds.). 1995. *Do ininteligivel ao sensivel*. São Paulo: Educ.
Lancioni, Tarcisio & Anna Maria Lorusso (eds.). 2020. Enunciazione e immagini. [Special Issue]. *E/C* 29. https://mimesisjournals.com/ojs/index.php/ec/index
Latour, Bruno. 1988. A Relativist Account of Einstein's Relativity. *Social Studies of Sciences*, 18. 3–44.
Latour, Bruno. 1993. *We Have Never Been Modern*. Cambridge (Mass.): Harvard University Press.
Latour, Bruno. 1999. *Pandora's Hope. Essays on the Reality of Science Studies*. Cambridge (Mass.), Harvard University Press.
Latour, Bruno. 2005. *Reassembling the Social*. Oxford: Oxford University Press.
Latour, Bruno. 2013. *An Inquiry into Modes of Existence: An Anthropology of the Moderns*. Cambridge, MA: Harvard University Press.
Latour, Bruno & Françoise Bastide. 1986. Writing science: Fact and fiction. In *Mapping the dynamics of science and technology*, Michel Callon, John Law, & Arie Rip (eds.). 51–66. London: Macmillan.
Latour, Bruno & Paolo Fabbri. 1977. La réthorique de la science. Pouvoir et devoir dans un article de science exacte. *Actes de la Recherche en Sciences Sociales* 13, 81–95.
Latour, Bruno & Steve Woolgar. 1979. *Laboratory Life: The Construction of Scientific Facts*. London: Sage.
Latour, Bruno & Peter Weibel (eds.). 2002. *Iconoclash. Beyond the Image Wars in Science, Religion, and Art*. Karlsruhe: ZKM.
Leach, Edmund. 1976. *Culture and Communication. The Logic by which Symbols are connected*. Cambridge (Mass.): Cambridge University Press.
Leone, Massimo. 2004. *Religious Conversion and Identity. The Semiotic Analysis of Text*. London: Routledge.
Leone, Massimo. 2010. *Saint and Signs. A Semiotic Reading of Conversion in Early Modern Catholicism*. Berlin-New York: De Gruyter.

Leone, Massimo (ed.). 2004. *La città come testo. Scritture e riscritture urbane.* [Special Issue]. *Lexia*, 1–2.
Leone, Massimo (ed.). 2009. *Attanti, attori, agenti. Senso dell'azione e azione del senso.* [Special Issue]. *Lexia*, 3–4.
Leone, Massimo (ed.). 2014. Efficacious Images. [Special Issue]. *Lexia* 17–18.
Levinson, Stephen C. 1983. *Pragmatics.* Cambridge University Press.
Lévi-Strauss, Claude. 1963 [1958]. *Structural Anthropology.* Chicago: Chicago University Press.
Lévi-Strauss, Claude. 1966 [1962]. *The Savage Mind.* Chicago: Chicago University Press.
Lévi-Strauss, Claude. 1978a [1968]. *The Origin of Table Manners. Mythologiques. Vol. 3*, Chicago: Chicago University Press.
Lévi-Strauss, Claude. 1978b. *Mith and Meaning.* Toronto University Press.
Lévi-Strauss, Claude. 1981 [1971]. *The Naked Man. Mythologiques. Vol. 4*, Chicago: Chicago University Press.
Lévi-Strauss, Claude. 1983 [1960]. *The Raw and the Cooked. Mythologiques. Vol. 1.* Chicago: Chicago University Press.
Lorusso, Anna Maria. 2015. *Perspectives on Semiotics of Culture.* New York: Palgrave-McMillan.
Lorusso, Anna Maria, Claudio Paolucci & Patrizia Violi (eds.). 2012. *Narratività. Problemi, analisi, prospettive.* Bologna: Bononia University Press.
Lotman, Yurij M. 1977. *The Structure of the Artistic Text.* Ann Arbor: University of Michigan.
Lotman, Yurij M. 1990. *Universe of the Mind: A Semiotic Theory of Culture.* London-New York: Tauris.
Lotman, Yurij M. 2009. *Culture and Explosion.* Berlin: Mouton-de Gruyter.
Lotman, Yurij, Boris A. Uspenskij, Vyacheslav Ivanov, Vladimir Toporov & Alexander M. Pjatigorskij. 1977. Theses on the semiotic study of cultures (as applied to Slavic texts). In Jan van der Eng & Mojmir Grygar (eds.) *Structure of Texts and Semiotics of Culture.* The Hague: Mouton.
Lotman, Yurij & Boris A. Uspensky. 1984. *The Semiotics of Russian Culture.* Ann Arbor: University of Michigan.
Lozano, Jorge. 1987. *El discurso histórico.* Madrid: Alianza.
Lozano, Jorge. 2012. *Persuasión. Estrategias del creer.* Bilbao: Universidad del Pais Vasco.
Lozano, Jorge, Cristina Peña-Marin & Gonzalo Abril. 1982. *Analisis del discurso.* Madrid: Catedra.
Lozano, Jorge & Daniele Salerno (eds.). Future. A Time of History. (Special Issue) *Versus* 132.
Lucid, Daniel P. (ed.). 1988. *Soviet Semiotics: An Anthology.* Baltimore: Johns Hopkins University Press. μ Group. 1970. *Rhétorique générale.* Paris: Larousse.
μ Group. 1991. *Traité du signe visuel.* Paris: Seuil
Manetti, Giovanni. 1996. *Knowledge through Signs. Ancient Semiotic Theories and Practices.* Turnhout: Brepols.
Manetti, Giovanni & Patrizia Violi. 1979. *L'analisi del discorso.* Milano: Espresso strumenti.
Mangano, Dario. 2011. *Semiotica e design.* Rome: Carocci.
Mangano, Dario. 2018. *Che cos'è la semiotica della fotografia.* Rome: Carocci.
Mangiapane, Francesco. 2018. *Retoriche social.* Palermo: Museo Pasqualino.
Mangiapane, Francesco & Tiziana M. Migliore (eds.). 2021. *Images of Europe. The Union between Federation and Separation.* Berlin: Springer.
Manovich, Lev. 2001. *The Language of New Media.* Cambridge: MIT Press.
Manovich, Lev. 2013. *Software Takes Command.* New York: Bloomsbury.

Marin, Louis. 1973. *Utopiques, jeux d'espaces*. Paris: Minuit.
Marin, Louis. 1986. *La parole mangée et autres essais théologico-politiques*. Paris: Klincksieck.
Marin, Louis. 1995 [1977]. *To Destroy Painting*. Chicago: Chicago University Press.
Marin, Louis. 2002 [2000]. *On Representation*. Stantford: Stantford University Press.
Marmo, Costantino, 1997. *Vestigia, imagines, verba. Semiotics and Logic in Medieval Theological Texts*. Turhout: Brepols.
Marrone, Gianfranco. 2001. *Corpi sociali. Processi comunicativi e semiotica del testo*. Turin: Einaudi.
Marrone, Gianfranco. 2007. *Il discorso di marca*. Rome-Bari: Laterza.
Marrone, Gianfranco. 2009 [2005]. *The Ludovico Cure. On Body and Music in 'A Clockwork Orange'*. Toronto: Legas.
Marrone, Gianfranco. 2014. L'age d'or de la sémiotique littéraire, et quelques consequences théoriques. *Signata* 5. 53–75.
Marrone, Gianfranco. 2016. *Semiotica del gusto*, Milan: Mimesis.
Marrone, Gianfranco. 2017a. *Sémiotique et critique de la culture*. Limoges: Pulim.
Marrone, Gianfranco. 2017b. La saisie esthétique, transformation non narrative de l'experience. In Thomas Broden & Stéphanie Walsh-Matthews (eds.). 2017b. La sémiotique post-greimassienne / Semiotics Post-Greimas. [Special Issue]. *Semiotica* 219. 115–132.
Marrone, Gianfranco. 2018. Narration et Conspiration. Formes de la bêtise dans le 'Pendule de Foucault'. *Cahiers de narratologie* 33. [En ligne]. http://journals.openedition.org/narratologie/7994.
Marrone, Gianfranco. 2019. Barthes, le récit, la saisie esthétique. In Veronica Estay Stange, Pauline Hachette & Raphaël Horrein (eds.). *Sens à l'horizon. Hommage à Denis Bertrand*. 39–48. Limoges: Lambert Lucas.
Marrone, Gianfranco (ed.). 1994. *Il testo filosofico*. Palermo: L'epos.
Marrone, Gianfranco (ed.). 2010. *Palermo. Esercizi di semiotica urbana*. Rome: Carocci.
Marrone, Gianfranco & Francesco Mazzucchelli (eds.). 2019. Forms of Life / Forms of the Body. [Special Issue]. *Versus* 128.
Marrone, Gianfranco (ed.). 2020. *Handbook of Culture and Communication of Taste*. https://www.cucota.eu/handbook-of-culture-and-communication-of-taste/
Marrone, Gianfranco & Dario Mangano. 2015. Brand language: methods and models of semiotic analysis. In Rossolatos, George (ed.). *Handbook of Brand Semiotics*. 46–88. Kassel: Kassel University Press.
Marrone, Gianfranco & Dario Mangano (eds.). 2018. *Semiotics of Animals in Culture*. Berlin: Springer.
Marsciani, Francesco. 2007. *Tracciati di etnosemiotica*. Bologne: Esculapio.
Marsciani, Francesco. 2012a. *Ricerche semiotiche. Vol. 1: Il tema trascendentale*. Bologne: Esculapio.
Marsciani, Francesco. 2012b. *Ricerche semiotiche. Vol. 2: In fondo al semiotico*. Bologne: Esculapio.
Marsciani, Francesco. 2019. La sémiotique générative de Greimas et sa valeur 'scientifique'. In Denis Bertrand, Jean-François Bordron, Ivan Darrault & Jacques Fontanille (eds.). *Greimas aujourd'hui: l'avenir de la structure*. 285–290. Paris: AFS éditions.
Martin, Browen & Felicitas Ringham. 2000. *Dictionary of Semiotics*. London-New York: Cassel.

Merleau-Ponty, Maurice. 1966 [1964]. *The Visible and the Invisible*. Evanston: Northwestern University Press.

Merleau-Ponty, Maurice. 2013 [1945]. *Phenomenology of Perception*. London: Routledge.

Metz, Christian. 1974. *Film Language: A Semiotics of the Cinema*. Chicago: Chicago University Press.

Metz, Christian. 2011 [1971]. *Language and cinema*. Berlin: The Gruyter.

Metz, Christian. 2016 [1991]. *Impersonal Enunciation, or the Place of Film*. New York: Columbia University Press.

Migliore, Tiziana. 2021. *De la part des usagers. Sémiotique de la reception*. Milan: Mimesis international.

Mitropoulou, Eleni. 2017. Le sens de l'echange. In Thomas Broden & Stéphanie Walsh-Matthews (eds.). 2017a. La sémiotique post-greimassienne / Semiotics Post-Greimas. [Special Issue]. *Semiotica* 214. 307–329.

Moreno Ruiz, Luisa. 2014. *Triptico en tono menor. Estudio semiotico*. Ediciones EyC

Moreno Ruiz, Luisa. Profondeur de la figure. In Veronica Estay Stange, Pauline Hachette & Raphaël Horrein (eds.). *Sens à l'horizon. Hommage à Denis Bertrand*. 63–76. Limoges: Lambert Lucas.

Morris, Charles. 1938. *Foundation of a Theory of Signs*. Chicago University Press.

Morris, Charles. 1971. *Writings on the General Theory of Signs*. The Hague: Mouton.

Nathan, Tobie. 2001. *Nous ne sommes pas seuls au monde*. Paris: Empêcheurs de penser en rond.

Nietzsche, Friedrich (1973). *Über Wahrheit und Lüge im aussermoralischen Sinne*; eng. transl. *On Truth and Lies in a Nonmoral Sense*. In *The Portable Nietzsche*, London: Viking Press, 1976.

Ochs, Elinor. 1988. *Culture and Language Development: Language Acquisition and Language Socialization in a Samoan Village*. Cambridge-New York: Cambridge University Press.

Ochs, Elinor. 2001. *Living Narrative: Creating Lives in Everyday Storytelling*. Cambridge (Mass.): Harvard University Press.

Ogden, Charlesv & Ivor Armstrong, Richards. 1923. *The Meaning of Meaning. A Study of the Influence of Language upon Thought and of the Science of Symbolism*. London: Routledge & Keagan Paul.

Oliveira de, Ana Claudia. 1997. *Vitrinas: acidentes estéticos na cotidianidade*. São Paulo: Educ.

Oliveira de, Ana Claudia (ed.). 2013. *As interações sensíveis. Ensaios de sociosemiótica a partir da obra de Eric Landowski*. São Paulo: Estaçao das Letras e Cores.

Oliveira de, Ana Claudia (ed.). 2014. *Do sensível ao inteligivel*. São Paulo: Estação das Letras e Cores.

Ong, Walter J. 1982. *Orality and Literacy. The Technologizing of the World*. London: Metheun.

Ouellet, Pierre. 1992. *Voir et savoir*. Montréal: Ed. Balzac.

Ouellet, Pierre. 2000. *Poétique du regard*. Limoges: Pulim.

Panier, Louis. 1984. *Récit et commentaires de la tentation de Jésus au désert: approche sémiotique*, Paris: Cerf.

Panier, Louis. 1991. *La naissance du Fil de Dieu*. Paris: Cerf.

Panofsky, Erwin. 1955. *Meaning in the Visual Art*. Garden City: Doubleday.

Parret Herman. 1986. *Les passions. Essai sur la mise en discours de la subjectivité*. Liège: Mardaga.

Parret, Herman. 1988. *Le sublime du quotidien*. Paris-Amsterdam: Hades-Benjamin.

Parret Herman. 1991. *Le sens et ses hétérogénéité*. Paris: CNRS.

Parret Herman. 2018. *Structurer. Progrès sémiotiques en épistemologie et en esthétique*. Louvain: Academia-L'Harmattan.
Parret Herman. 2019. Transparence et opacité du discours. In Veronica Estay Stange, Pauline Hachette & Raphaël Horrein (eds.). *Sens à l'horizon. Hommage à Denis Bertrand*. 15–26. Limoges: Lambert Lucas.
Parret, Herman & Hans-George Ruprecht (eds.). 1985. *Exigences et perspectives de la sémiotique / Aims and Prospects of Semiotics*. 2 vols. Amsterdam-Philadelphia: John Benjamins
Patte, Daniel. 1990. *The Religious Dimensions of Biblical Texts*. Atlanta: Scholar Press.
Pavel, Thomas. 2017. Human action in narrative grammars. In Thomas Broden & Stéphanie Walsh-Matthews (eds.). 2017a. La sémiotique post-greimassienne / Semiotics Post-Greimas. [Special Issue]. *Semiotica* 214. 219–229.
Pavis, Patrice & Rodrigue Villeneuve (eds.). 1993. Gestualités. [Special Issue]. *Protée* 21, 3.
Peirce, Charles S. 1931–58. *Collected Papers*. 8 vols. Cambridge (Mass.): Harvard University Press.
Perron, Paul. 2003. *Narratology and Text*. Toronto: University of Toronto Press.
Perron, Paul & Marcel Danesi. 1993. *A. J. Greimas and Narrative Cognition*. Toronto: Toronto Semiotic Circle.
Perron, Paul & Franck Collins (eds.). 1989. *Paris School Semiotics*. 2 vols. Amsterdam-Philadelphia: John Benjamins.
Petitot, Jean. 1985. *Morphogenèse du sens*. Paris: Puf.
Petitot, Jean. 2019. Phénoménologie de la structure: de l'idéalité formelle à la structure cognitive. In Denis Bertrand, Jean-François Bordron, Ivan Darrault & Jacques Fontanille (eds.). *Greimas aujourd'hui: l'avenir de la structure*. 13–24. Paris: AFS éditions.
Petitot, Jean & Paolo Fabbri (eds.). 2000. *Au Nom du Sens. Autour de l'oeuvre d'Umberto Eco*. Paris: Grasset.
Petöfi, Janos S. (ed.). 1983. Methodological aspects of discourse processing. [Special Issue]. *Text. An interdisciplinary journal for the study of discourse*, 3, 1.
Peverini, Paolo & Marica Spalletta. 2009. *Unconventional*. Rome: Meltemi.
Pezzini, Isabella. 1998. *Le passioni del lettore*. Milan: Bompiani.
Pezzini, Isabella. 2001. *Lo spot elettorale*. Rome: Meltemi.
Pezzini Isabella (ed.). 1991. *Semiotica delle passioni*. Bologna: Esculapio.
Pezzini Isabella (ed.). 2001. *Semiotic Efficacity and the Effectiveness of the Text. From Effects to Affects*. Turnhout: Brepols.
Pezzini Isabella (ed.). 2002. *Trailer, spot, clip, siti, banner. Le forme brevi della comunicazione audiovisiva*. Rome: Meltemi.
Pezzini Isabella (ed.). 2009. *Roma: luoghi del consumo, consumo dei luoghi*. Rome: Nuova Cultura.
Pezzini Isabella (ed.). 2016. *Roma in divenire, tra identità e conflitti*. Rome: Nuova Cultura.
Pezzini Isabella & Vincenza Del Marco (ed.). 2012. *Passioni collettive. Cultura, politica e società*. Rome: Nuova Cultura.
Paolucci, Claudio. 2010. *Strutturalismo e interpretazione*. Milan: Bompiani.
Paolucci, Claudio. 2020. *Persona. Soggettività nel linguaggio e semiotica dell'enunciazione*. Milan: Bompiani.
Polidoro, Piero (ed.). 2018. Fake News, Misinformation/ Disinformation,Post-Truth. [Special Issue]. *Versus* 125.
Pozzato Maria Pia. 2001. *Semiotica del testo*. Rome: Carocci.

Pozzato Maria Pia. 2012. *Foto di matrimonio e altri saggi*. Milan: Bompiani.
Prieto, Louis. 1966. *Messages et signaux*. Paris: Puf.
Prieto, Louis. 1975. *Pertinence et pratique. Essai de semiologie*. Paris: Minuit.
Propp, Vladimir Ja. 1958 [1928]. *Morfology of the Folktale*. Bloomington: Indiana University Press.
Quéré, Henri. 1992. *Intermittences du sens*. Paris: Puf.
Rabatel, Alain. 2017. L'énonciation, la praxis énonciative et le discours. In Thomas Broden & Stéphanie Walsh-Matthews (eds.). 2017b. La sémiotique post-greimassienne / Semiotics Post-Greimas. [Special Issue]. *Semiotica* 219. 273–291.
Rastier, François. 1987. *Sémantique interpretative*. Paris: Puf.
Rastier, François. 2001. *Arts et sciences du texte*. Paris: Puf.
Rastier, François. 2015. *Saussure au futur*. Paris: Les Belles-Lettres/Encre Marine.
Rastier, François. 2019. Greimas et la linguistique. In Denis Bertrand, Jean-François Bordron, Ivan Darrault & Jacques Fontanille (eds.). *Greimas aujourd'hui: l'avenir de la structure*. 202–213. Paris: AFS éditions.
Rector, Monica. 1979. *Para ler Greimas*. Rio de Janeiro: Francisco Alves.
Rector, Monica (ed.). 2002. Los gestos. Sentidos y practices. [Special Issue]. *deSignis* 3.
Ricci, Piero. 1994. *Nomi Pieghe Tracce. Studi di semiotica della cultura*. Urbino: Quattroventi.
Ricoeur, Paul. 1973. The Model of the Text: Meaningful Action considered as Text. *New Literary History*, 5. 91–117.
Ricoeur, Paul. 1974 [1969]. *The Conflict of Interpretations: Essays in Hermeneutics*. Evanston: Northwestern University Press.
Ricoeur, Paul. 1984–1988 [1983–1985]. *Time and Narrative*. 3 vols. Chicago: Chicago University Press.
Ricœur, Paul. 1986. *Du texte à l'action. Essais d'herméneutique*. Paris: Seuil.
Rieusset-Lemarié, Isabelle. 2019. Autonomie des 'sujets de faire' dans les dispositifs modaux et ouverts (de la *Sémiotique des passions* à l'esthétique de l'inattendu). In Denis Bertrand, Jean-François Bordron, Ivan Darrault & Jacques Fontanille (eds.). *Greimas aujourd'hui: l'avenir de la structure*. 350–367. Paris: AFS éditions.
Riffaterre, Michael. 1971. *Essais de stylistique structurale*. Paris: Flammarion.
Riffaterre, Michael. 1979, *La production du texte*. Paris: Seuil.
Ritchie, Mary Key. 1975. *Paralanguage and Kinesics (Nonverbal Communication)*. Metuchen (N.J.): The Scarecrow Press.
Rorty, Richard. 1979. *Philosophy and the Mirror of Nature*. Princeton University Press.
Saint-Martin. Fernand. 1990. *Semiotics of Visual Language*. Bloomington: Indiana University Press.
Saussure de, Ferdinand. 1988 [1917]. *Course in General Linguistic*. London: Open Court.
Saussure de, Ferdinand. 2002. *Ecrits de linguistique générale*. Simon Bouquet & Rudolf Engler (eds.). Paris: Gallimard.
Schapiro, Micheal (ed.). 1984. *Language and Politics*. Oxford: Basil Blackwell Publisher.
Sebeok, Thomas A. 1976. *Contributions to the Doctrine of Signs*. Lanham: America University Press.
Sebeok, Thomas A. 1979. *The Signs and Its Masters*. Austin: Texas University Press.
Sebeok, Thomas A. 1986. *I Think I Am a Verb. More Contributions to the Doctrine of Signs*. New York: Plenum.
Sebeok, Thomas A. 1991. *A Sign is Just a Sign*. Bloomington: Indiana University Press.
Sedda, Franciscu. 2012 *Imperfette traduzioni*. Rome: Nuova Cultura.
Sedda, Franciscu. 2019. *Tradurre la tradizione*. Milan: Mimesis.

Sedda, Franciscu & Paolo Sorrentino. 2019; *Roma. Piccola storia simbolica*. Rome: La lepre.
Sedda, Franciscu & Paolo Sorrentino (eds.). 2020. *Islandness. Cultural semiotics of Island*. [Special Issue]. *Lexia*, 35–36.
Segre, Cesare. 1979. *Semiotica filologica*. Turin: Einaudi.
Semprini, Andrea. 1995. *L'objet comme procès et comme action*. Paris: L'Harmattan.
Serra, Marcello (ed.). 2012. *En torno a la semiótica de la cultura*. Madrid: Fragua.
Shapiro, Meyer. 1973. *Words and Pictures. On the Literal and the Symbolic in the Illustration of a Text*. The Hague and Paris: Mouton.
Sherzer, Joel. 1983. *Kuna Ways of Speaking*. Austin: University of Texas Press
Shklovsky, Viktor. 1990 [1925]. *Theory of Prose*. Champaign: Dalkey Archive Press.
Sonesson, Göran. 2017. Greimasean Phenomenology and beyond: From Isotopy to Time Consciousness. In Broden & Welsh (eds.) 2017b. 93–113.
Spaziante, Lucio. 2007. *Sociosemiotica del pop*. Rome: Carocci.
Stano, Simona (ed.). 2015. Food and Cultural Identity. [Special Issue]. *Lexia*, 19–20.
Stano, Simona (ed.). 2016. Semiotics of Food. [Special Issue]. *Semiotica*, 211.
Steiner, Geroge. 1975. *After Babel*. New York-London: Oxford University Press.
Stoichita, Victor. 1989. *L'instauration du tableau: métapeinture à l'aube des temps modernes*. Genève: Droz.
Stoichita, Victor. 1997a. *A Short History of the Shadow*. London: Reaktion Books.
Stoichita, Victor. 1997b. *The Self-Aware Image. An insight into Early Modern Meta-Painting*. New York: Cambridge University Press.
Tarasti, Eero. 1979. *Myth and Music: A Semiotic Approach to the Aesthetics of Myth in Music*. Berlin: De Gruyter.
Tarasti, Eero. 1994. *A Theory of Musical Semiotics*. Bloomington: Indiana University Press.
Thom, René. 1988. *Esquisse d'une sémio-physique*. Paris: Interéditions.
Thom, René. 1990. *Apologie du logos*. Paris: Hachette.
Todorov, Tzvetan. 1973 [1970]. *The Fantastic; a structural approach to a literary genre*. Cleveland: Press of Case Western Reserve University.
Todorov, Tzvetan. 1977 [1971]. *The Poetics of Prose*. Oxford: Blackwell.
Todorov, Tzvetan. 1982 [1978]. *Symbolism and interpretation*. Ithaca (N.Y.): Cornell University Press.
Todorov, Tzvetan (ed.). 1966. *Théorie de la littérature. Textes des formalistes russes*. Paris: Seuil.
Tomačevskij, Boris. 1928. *Teorija literatury. Poetika* [Literary Theory. Poetics]. Leningrad: Gosudarstvennoe izdatel'stvo [State Publishers].
Thibault, Paul J. 1991. *Social Semiotics as Praxis*. University of Minnesota Press.
Thürlemann, Félix. 1982. *Paul Klee. Analyse sémiotique de trois peintures*. Lausanne: Age d'homme.
Torop, Peter. 1995. *Total'nyj perevod* [The Total Translation]. Tartu: Tartu University Press.
Torop, Peter, Mihhail Lotman & Kalevi Kull (eds.). 2006. Between semiotics and anthropology: Life histories and other methodological issues. [Special Issue]. *Sign Systems and Studies* 34, 2.
Traini, Stefano. 2006. *Le due vie della semiotica*. Milan: Bompiani.
Traini, Stefano. 2008. *Semiotica della comunicazione pubblicitaria*. Milan: Bompiani.
Trubeckoj, Nicolaj S. 1939. *Grundzuge der Phonologie*. Prague: Ed. du Cercle de Prague.
Umiker-Sebeok, Jean (ed.). 1987. *Marketing and Semiotics. New Directions in the Study of Signs for Sale*. Berlin-New York-Amsterdam: Mouton-de Gruyter.

Uspenskij Boris. 1973. *A Poetics of Composition*. Berkeley: California University Press.
Uspenskij, B. A., Ivanov, V.V., Toporov, V.N., Pjatigorskij A.M., Lotman, Ju. M. 2003. Theses on the semiotic study of cultures (as applied to Slavic texts). In: Gottdiener, Mark, Boklund-Logopoulou, Karin & Lagopoulos, Alexandros Ph. (eds.). *Semiotics. Vol. 1*. London: Sage. 293–316.
Van Dijk, Theun A. 1972. *Some Aspects of Text Grammars. A Study in Theoretical Linguistics and Poetics*. The Hague-Paris: Mouton.
Dijk van, Theun. 1977. *Text and Context*. London: Longman.
Van Dijk, Theun A. (ed.). 1985. *Handbook of Discourse Analysis*. 4 vols. London: Academic Press.
Ventura Bordenca, Ilaria. 2020. *Essere a dieta*. Milan: Meltemi.
Verón, Eliseo. 1996. *Conducta, estructura y comunicación*, Buenos Aires: Amorrortu.
Vibaek-Pasqualino, Janne (ed.). 1979. *Strutture e generi delle letterature etniche*. Palermo: Flaccovio.
Violi, Patrizia. 2001 [1997]. *Meaning and experience*. Bloomington: Indiana University Press.
Violi, Patrizia. 2017 [2014]. *Landscapes of Memory. Trauma, Space, History*. Oxford-Bern-Berlin-Bruxelles-Frankfurt am Main-New York-Wien: Peter Lang.
Violi, Patrizia. 2019. Les vois des autres. In Veronica Estay Stange, Pauline Hachette & Raphaël Horrein (eds.). *Sens à l'horizon. Hommage à Denis Bertrand*. 299–308. Limoges: Lambert Lucas.
Viveiros De Castro, Eduardo. 2009. *Métaphisiques cannibales. Lignes d'antropologie post-structurale*. Paris: Puf.
Volli, Ugo. 1997. *Fascino*. Milan: Feltrinelli.
Volli, Ugo. 2002. *Semiotik. Eine Einfürung in irhe Grundbegriffe*. Berlin: A. Franke.
Wagner, Roy. 1975. *The Invention of Culture*. University of Chicago.
Wahl, François. 1972. Texte. In Oswald Ducrot & Tzvetan Todorov (eds.). *Dictionnaire encyclopédique des sciences du langage*. Paris: Seuil. 306–321
Wittgenstein, Ludwig. 1953. *Philosophische Untersuchungen*. Oxford: Basil Blackwell.
Wöfflin, Heinrich. 1915. *Kunstgeschichtliche Grundbegriffe. Das Problem der Stilentwicklung in der neueren Kunst*. München: Schwabe & Co.
Zilberberg, Claude. 1988. *Raison et poétique du sens*. Paris: Puf.
Zilberberg, Claude. 1982, *Essai sur les modalités tensives*. Amsterdam: John Benjamins.
Zilberberg, Claude. 2006. *Eléments de grammaire tensive*. Limoges: Pulim.
Zilberberg, Claude. 2011. *Des formes de vie aux valeurs*. Paris: Puf.
Zinna, Alessandro. 2017. Hjelmslev, la sémiotique et l'Ecole de Paris. In Thomas Broden & Stéphanie Walsh-Matthews (eds.). 2017b. La sémiotique post-greimassienne / Semiotics Post-Greimas. [Special Issue]. *Semiotica* 219. 455–470.
Zinna, Alessandro. 2019. L'autonomie du plastique et du figuratif. Le dispositif de l'énonciation de l'Icone. In Veronica Estay Stange, Pauline Hachette & Raphaël Horrein (eds.). *Sens à l'horizon. Hommage à Denis Bertrand*. 127–142. Limoges: Lambert Lucas.
Zinna, Alessandro (ed.). 1997. *Hjelmslev aujourd'hui*. Turnhout, Brepols.
Zinna, Alessandro & Lorenzo Cigana (eds). 2017. *Louis Hjelmslev (1899–1965). Le forme del linguaggio e del pensiero*. Toulouse: Editions CAMS/O.
Zunzunegui, Santos. 2003. *Metamorfosis de la mirada. Museo y semiotica*. Madrid: Catedra.
Zunzunegui, Santos. 2005. *Las cosas de la vida. Lecciones de semiotica estructural*. Madrid: Biblioteca Nueva.

Index

abduction/deduction/induction 4–5
actant/actor 18, 23–24, 39–40, 43–44, 46–47, 51, 55, 57, 67, 69, 72–73, 79, 152
actual (subject) 41
addresser 43–48, 54, 68–69, 160, 171
adiaphoria 36, 55
aesthetic grasp 111–114, 116
aesthetics / aesthesis 36, 60, 101, 111, 130, 153, 156
antagonist 45, 47, 49–50
anti-subject 45, 49–51, 70
aspectuality 54
assistant 73
– participant 73
– protagonist 73
axiology 23, 35–36, 101
– of consumption values 161–162

biplanarity 16–17, 123
body 2, 54, 56, 77, 84–85, 87, 91, 94, 98–107, 112–113, 116, 118, 126, 134, 173

canonical schema of passions 55–57
category
– chromatic 97
– eidetic 97
– semantic 29, 32, 34–35, 60
– topologic 97
classic/baroque 98–99
coherence 17, 76–78, 91,129
communication/signification 3–4, 54, 56, 62–63, 69–70, 107, 110, 122
communicative pact 41, 63, 70, 77
competence 45–47, 49, 57, 59, 147–148
complementarity 23, 28–32
configuration
– thematic 91
– textual 15, 82, 142
conjunction/disjunction 37–38, 40–42, 50, 53, 68, 86, 117–118
consistency 15, 17, 85
constitution 55–57, 108
content (plane of) 79, 81, 84–85, 90, 102

context 15, 34–35, 40, 63, 94, 123–124, 129, 137, 140–141, 147, 150, 159–160, 168–169, 171–172
contract 46–47, 52
contraction/expansion 60–61
contradiction (privative opposition) 23, 28–31, 33
contrariety (qualitative opposition) 29–30
cultural models 166–172

de-realization 74
deconstructionism 136–140
deduction See abduction/deduction/induction
design 10, 14–15, 25, 27, 62, 75, 84, 121, 123, 125, 145, 159–160
difference 6–8
dimension
– cognitive 47, 51, 72, 109, 173
– strategic 51–53, 58, 70, 76, 108
– passionate 38, 57, 129
– pragmatic 56, 101
– somatic (aesthetic) 56, 99, 101, 103, 111
discourse 8, 15, 23, 25, 27, 35, 54, 62–63, 70–72, 74–76, 79–82, 102, 123–124, 130, 132–133, 136, 139, 141, 150–151, 153, 159–161
disposition 53–56
donor 44, 46

effect of the real 82
efficacy/efficiency 62, 70–71, 75
ellipsis 48
emotion 5, 55–56, 108
empirical level 155–156
encyclopedia 19, 147–150
enunciation VI 18, 23, 62–68, 70–73, 75, 79, 97, 102, 103, 109, 129, 133, 139–144, 174
enunciator/enunciatee 18, 27, 68–73, 75–77
etic/emic 167
expression (plane of) 1–2, 8–9, 11, 16–17, 75, 81, 84, 87–88, 90, 99, 102, 112, 134, 136, 145–146, 150, 155, 168, 170

https://doi.org/10.1515/9783110688986-008

fashion 15, 98–99, 122–123, 126, 144, 159–160
figural 81–82, 94–95, 105–106
figurative reasoning 91–95
figurative syntax 103–105
– of smell 105
figurative/plastic 85, 87–88, 90, 94–98, 103, 105–106, 112, 153
figurative/thematic 78–82, 91, 102
figurativity 78, 81–82, 88, 91–94, 102, 112
focaliser 73
form
– of expression 16
– of content 16
form/substance 9–11, 16
forms of life 42, 58–61, 115, 125, 162

generative path VI, 18, 22–24, 28, 79–80, 150–158
genre (of discourse) 149–159

having-to 41–42, 44, 46, 50, 54, 69
hermeneutic arc 140–142
hermeneutics 135–136, 140–142, 142–144

iconic (level) 82, 94–95
identity 40–42, 50, 53, 77
image 9, 18, 51–52, 70–71, 84–90, 91–96
induction See abduction/deduction/induction
inference 4–6
informer 72–75
inter-objectivity 75
inter-subjectivity 112, 140
inter-textuality 18–19, 75–78, 131, 141
interdiscursivity 75–78
intermediality 75–78
interpretative
– Cooperation 146–148
– anthropology 165–166

langue/parole 64, 66, 144–145
levels of meaning 21, 84, 166
linguistic act 66

manipulation 46–48, 53, 57, 59, 108, 161, 173

modalities 40–42, 44, 79, 102, 109
– virtualising 41
– actualising 41
moralisation 55, 57, 108
myth 163–164

narrative grammar 29, 37, 108, 113, 132, 151
narrative programme 23, 28, 40, 42–43, 45, 49, 51–53, 59, 77
– basic 42–43
– instrumental 42–43
narrative schema 45–49, 55, 57–59, 108
narrativity 27–28, 37–38, 45, 51, 55, 58, 67, 72, 108, 123, 132–133
– narration 22, 27, 48–49

object of value 38–43, 50, 54, 59, 72, 86
observer 72–75
outside-text 125, 136–140, 152–155

passion 7, 11, 41–42, 53–58, 108–109, 111, 116, 148
– without a name 54–56
pathemisation 55–56, 108
performance 45–51, 56–57, 59, 64, 115
performative utterances 68
pertinence VI 13, 15, 19–20, 22, 24, 48
philology and text criticism 127–128
plastic (semiotics) 85–86, 88–90, 96–97, 113
point of view 50, 72, 74, 76, 107, 123, 164, 173
polemics (narrative) 49, 52, 108
presupposition 48–49, 75, 106, 109, 145, 170, 174
– reciprocal 107, 110, 127, 149, 172
process
– communicative 52, 63, 69
– narrative 27, 32–33, 88, 41
– of signification 107
– pathemic 53–55
processuality 122–123, 132
public opinion 161

realised (subject) 41
realism 67, 74, 82, 92, 94, 172
– integral 74–75

– objective 74
– subjective 74
receiver See sender/receiver relationship
– communicative 69
referential 69
representation (farewell to) 172–174

sanction 46–49, 57, 59, 108
science of the concrete 164
semes 34–35
– figurative 94
semiosphere VI 15, 38, 124, 136, 170, 172
semiotic square VI 23, 28–37, 74, 79, 110, 151–152
– operations of 28, 32–33, 37
– relationships of 23, 29–33, 36
– second generation terms 34
– timic category 34, 36–38
semiotics
– general 161
– of culture 162–171
– socio VI 15, 121–124, 124–127, 135–136, 158–162
semi-symbolism 90, 97
sender/receiver 2–3
sensitization 55–57
shifting-in 67
shifting-out 67–68, 75
signification See communication/signification
source 63, 68–71, 104–106
space 23–25, 66–67, 73, 75–78, 85–87, 92, 97, 99–102, 105–110, 152, 167, 170
spectator 60, 72–73, 87, 94–95
strategies/tactics 51–53, 58, 70, 108
structural anthropology 165–166
structuralism/structuralist 13, 17, 64, 121, 126, 134–135, 137, 170
structures
– discursive 23, 39, 79
– elementary (of signification) 28–36, 84, 152
– narrative VI 11, 22–23, 28, 37–49, 79, 151
Subject 3, 9, 18, 38–51, 53–54, 56–60, 62–63, 65–72, 86–87, 101, 107, 109–110, 112–116, 152

– object 37–42, 44, 49–50, 58–59, 68, 79, 86, 162, 171
– of being 38, 40–43, 68–69
– of doing 38–43, 68–69
subjectivity/objectivity 74–75, 116, 147
synaesthesia 102–106, 116

target 18, 40, 63, 104–106
task
– decisive 45
– glorious 47
– qualyfing 46
tension 1, 32, 36, 51, 54, 60, 148, 166, 168
term
– neutral 34
– complex 34
textual closure 17, 19, 154
textual linguistics 128–129
textualisation VI 23, 84, 144–146, 156–158
thematic/figurative See figurative/thematic
theme/thematic 63, 80, 81
thick description 166
timic 35–36, 38, 54, 79, 88, 97
totality
– integral 77
– partitive 77
translation (inter-semiotic) 77
trust 27, 46, 62, 71
truth 51, 68–69, 74, 82, 114, 127, 137, 140–141, 158

valorisation 14, 33, 35–36, 39, 53, 161
value 5, 6–7, 8, 9, 11, 20, 23, 28, 33, 35–36, 37, 38–39, 40, 42–43, 45, 46, 49–50, 53, 70, 79, 86, 101, 129, 132–133, 140, 141, 158, 161
– basic 43
– instrumental 42–43, 161
value of value 44, 53
virtual (subject) 41–43, 50

work/text 130–132
worth 44, 45

www.ingramcontent.com/pod-product-compliance
Lightning Source LLC
Chambersburg PA
CBHW050526170426
43201CB00013B/2096